The Handbook of Marketing Strategy for Life Sciences Companies

The Handbook of Marketing Strategy for Life Sciences Companies
Formulating the Roadmap You Need to Navigate the Market

Jean-Francois Denault

A PRODUCTIVITY PRESS BOOK

Routledge
Taylor & Francis Group
711 Third Avenue, New York, NY 10017

© 2018 by Taylor & Francis Group, LLC
Productivity Press is an imprint of Taylor & Francis Group, an Informa business

No claim to original U.S. Government works

Printed on acid-free paper

International Standard Book Number-13: 978-0-8153-7690-3 (Hardback)
International Standard Book Number-13: 978-0-8153-7688-0 (Paperback)

This book contains information obtained from authentic and highly regarded sources. Reasonable efforts have been made to publish reliable data and information, but the author and publisher cannot assume responsibility for the validity of all materials or the consequences of their use. The authors and publishers have attempted to trace the copyright holders of all material reproduced in this publication and apologize to copyright holders if permission to publish in this form has not been obtained. If any copyright material has not been acknowledged please write and let us know so we may rectify in any future reprint.

Except as permitted under U.S. Copyright Law, no part of this book may be reprinted, reproduced, transmitted, or utilized in any form by any electronic, mechanical, or other means, now known or hereafter invented, including photocopying, microfilming, and recording, or in any information storage or retrieval system, without written permission from the publishers.

For permission to photocopy or use material electronically from this work, please access www.copyright.com (http://www.copyright.com/) or contact the Copyright Clearance Center, Inc. (CCC), 222 Rosewood Drive, Danvers, MA 01923, 978-750-8400. CCC is a not-for-profit organization that provides licenses and registration for a variety of users. For organizations that have been granted a photocopy license by the CCC, a separate system of payment has been arranged.

Trademark Notice: Product or corporate names may be trademarks or registered trademarks, and are used only for identification and explanation without intent to infringe.

Visit the Taylor & Francis Web site at
http://www.taylorandfrancis.com

and the Productivity Press site at
http://www.ProductivityPress.com

Contents

List of Figures xiii
List of Tables xv
Preface xvii
Introduction to Marketing Strategy xxi
Author xxix

1　Marketing Strategy Road Map 1
　1.1　Planning toward Strategy 1
　1.2　Planning Your Marketing Strategy 2
　　　1.2.1　Performing Market Research 2
　　　　　　1.2.1.1　Planning Your Data Collection 3
　　　　　　1.2.1.2　Overview of Market Research Tools 3
　　　1.2.2　Situation Analysis 4
　　　　　　1.2.2.1　Internal Analysis 4
　　　　　　1.2.2.2　External Analysis 4
　　　1.2.3　Developing a Marketing Strategy 5
　　　1.2.4　Implementation and Control Mechanisms 5
　　　1.2.5　Final Notes 6

2　Overview of Market Research 7
　2.1　Basic Market Research Concepts 8
　　　2.1.1　Primary and Secondary Market Research 8
　　　2.1.2　Quantitative and Qualitative Data 8
　　　2.1.3　Miles-Wide versus Deep Dive Research 10
　2.2　Preparing Your Market Research Plan 11

v

2.3 Collecting Data—Primary Research 12
  2.3.1 Data Collection Methods 13
   2.3.1.1 In-Depth Interviews 13
   2.3.1.2 Focus Groups 14
   2.3.1.3 Online Surveys 16
 2.4 Secondary Research 17
  2.4.1 Active Secondary Research...............................18
   2.4.1.1 Popular Sources of Data Online 19
   2.4.1.2 Using Search Engines to Look
           for Information 20
  2.4.2 Passive Secondary Research 22
  2.4.3 Internal Secondary Data 24
 2.5 A Few Words on Ethics and Market Research 25
 References 26

3 Situation Analysis 29
 3.1 Internal Analysis...30
  3.1.1 Assessing Your Corporate Vision and
        Mission Objectives 31
  3.1.2 Assessing Your Current Capabilities...............32
  3.1.3 Assessing Your Company's Business Model.....34
  3.1.4 Some Final Notes on Internal Situation
        Analysis 39
 3.2 External Analysis..40
  3.2.1 Customer Analysis 40
   3.2.1.1 Understanding the Customer's
           Decision-Making Process 41
   3.2.1.2 The Client Ecosystem........................ 44
   3.2.1.3 What Does My Customer
           Want?—Using the Kano Model to
           Understand Your Customer................50
   3.2.1.4 Identifying Customer Behavior 52
   3.2.1.5 Building Customer Profiles 54
   3.2.1.6 Some Final Notes on Customer
           Behavior..55
  3.2.2 Competitor Analysis...57

		3.2.2.1 Preparation 58
		3.2.2.2 Identify Key Competitors 61
		3.2.2.3 Evaluate Your Competitors................ 61
		3.2.2.4 Getting Information on Competition..................................... 66
		3.2.2.5 Final Notes on Competition.............. 68
	3.2.3	Market Analysis.. 68
		3.2.3.1 Market Size Estimation—The TAM-SAM-SOM Model 68
		3.2.3.2 Market Forecasting 73
		3.2.3.3 Final Notes on Market Analysis 76
	3.2.4	Environmental Analysis 76
		3.2.4.1 Microenvironment—Porter's Five Forces.. 77
		3.2.4.2 Macro-Environment—The SLEPT Model.. 80
		3.2.4.3 Building a SLEPT Model 82
3.3	Classifying Outputs: From SWOT to TOWS 85
	3.3.1	The SWOT Model 85
		3.3.1.1 The Four Elements of a SWOT Model.. 86
		3.3.1.2 Developing Strategy Applications— From SWOT to TOWS 88
3.4	Concluding Remarks 90
References 90

4 Developing a Marketing Strategy 91

4.1	Selecting Your Marketing Strategy Vision................. 91
	4.1.1 Strategic Commitment..................................... 92
	4.1.2 Strategic Opportunism..................................... 93
	4.1.3 Strategic Adaptability 95
4.2	Choosing Your Marketing Model—What Type of Company Are You?... 95
	4.2.1 The Classical Company Models........................ 96
		4.2.1.1 The Production Model 96
		4.2.1.2 The Product-Focused Model 97

    4.2.1.3 The Selling Model 97
    4.2.1.4 The Marketing Model 98
  4.2.2 Modern Marketing Concepts 99
    4.2.2.1 Relationship Marketing 99
    4.2.2.2 Integrated Marketing 100
    4.2.2.3 Internal Marketing 100
4.3 Creating, Adapting, and Implementing Strategy 101
  4.3.1 Determining Your Target Market—
    Segmentation 101
  4.3.2 Strategies Based on Products and Market 102
    4.3.2.1 Market Penetration 103
    4.3.2.2 Market Development 105
    4.3.2.3 Product Development Strategies 107
    4.3.2.4 Product Diversification 108
  4.3.3 Marketing Strategies Based on
    Competitive Advantage 110
    4.3.3.1 Cost Leadership 111
    4.3.3.2 Differentiation 112
    4.3.3.3 Cost and Differentiation Focus 114
4.4 Developing the Marketing Mix 114
  4.4.1 Product 115
    4.4.1.1 Key Product Decisions 115
    4.4.1.2 Measuring Product-Market Fit 116
    4.4.1.3 Making Key Product Decision—
      The ICE Score 117
    4.4.1.4 Building a Better Product—The
      Hook Model 119
  4.4.2 Pricing Strategy 121
    4.4.2.1 Cost-Based Pricing 121
    4.4.2.2 Competitive-Based Pricing 121
    4.4.2.3 Customer Value-Based Pricing 122
    4.4.2.4 Price Skimming 122
    4.4.2.5 Freemium Pricing 123
    4.4.2.6 Other Considerations for Your
      Pricing 123

		4.4.3	Promotion Strategy 126
			4.4.3.1 Objectives of Your Promotion Strategy 126
			4.4.3.2 Promotional Tools 127
			4.4.3.3 Choosing Your Promotional Message and Channel 129
			4.4.3.4 Choosing Your Promotional Tools: The Bullseye Framework 130
		4.4.4	Distribution Strategy 132
			4.4.4.1 Factors to Choose Your Distribution Strategy.........................133
			4.4.4.2 Different Types of Distributions 134
			4.4.4.3 Specific Considerations for Distribution in Life Sciences 138
	4.5	The Role of Digital Marketing................................138
		4.5.1	Driving Traffic 139
		4.5.2	Selling Products Online 140
		4.5.3	Digital Marketing in Action—The Pirate Metrics: "AARRR!" 141
	References 142

5 Marketing Strategy Implementation and Control... 145

	5.1	Implementation..146
		5.1.1 Implementation versus Strategy 146
		5.1.2 Implementation Plan......................................147
		5.1.3 Barriers to Successful Implementation of Marketing Strategy 149
		5.1.3.1 External Pressures of the Organization......................................149
		5.1.3.2 Internal Pressures of the Marketing Function 150
	5.2	Control Elements 151
		5.2.1 Implementation of Control Processes 152
		5.2.2 Barriers to the Successful Implementation of Control Procedures...................................153
		5.2.2.1 Inadequate Monitoring......................153

	5.2.2.2	Inadequate Targets	153
	5.2.2.3	Management by Exceptions	153
	5.2.2.4	Cost and Complexity	153
5.2.3	A Word of Caution on Control Systems		154

6 Marketing Metrics ... 155
- 6.1 Why Use Metrics? ... 156
- 6.2 Some Pre-revenue Ratios ... 156
  - 6.2.1 Sales Force Coverage ... 156
  - 6.2.2 Break-Even Analysis ... 157
- 6.3 Ratios to Measure Sales Effectiveness ... 160
  - 6.3.1 Return on Sales ... 160
  - 6.3.2 Advertising-to-Sales Ratio ... 160
  - 6.3.3 Customer Acquisition Cost ... 161
  - 6.3.4 Marketing Percentage of CAC ... 161
  - 6.3.5 Average Retention Cost ... 162
  - 6.3.6 Lifetime Customer Value ... 162
- 6.4 Digital Marketing Metrics ... 164
  - 6.4.1 Traffic Metrics ... 164
    - 6.4.1.1 Overall Site Traffic ... 165
    - 6.4.1.2 Monitoring the Source of Web Traffic ... 165
    - 6.4.1.3 Monitoring the Paid Traffic ... 166
  - 6.4.2 Conversion Metrics ... 166
  - 6.4.3 Revenue Metrics ... 168
- 6.5 Final Notes ... 168

7 Discussion on Unique Perspectives ... 171
- 7.1 Marketing in Life Sciences ... 171
- 7.2 Marketing Health-Care Services ... 174
  - 7.2.1 Difference between Services and Products ... 174
  - 7.2.2 Developing Marketing Strategies for Health-Care Services ... 175
- 7.3 Marketing Health-Care Digital Products ... 179
  - 7.3.1 Developing Marketing Strategies for Digital Products ... 180
  - 7.3.2 Issues with Health-Care Digital Products ... 182

7.4 Final Notes	184
Reference	184

8 Final Thoughts 185

Bibliography and Further Reading 189

Index 193

List of Figures

Figure 1.1 Basics of the market research process 3

Figure 3.1 Major steps toward commercialization in therapeutics 34

Figure 3.2 The classical customer purchasing process 42

Figure 3.3 Example of a customer persona: Terry—The typical lab tech 56

Figure 3.4 TAM-SAM-SOM model framework 70

Figure 3.5 TAM-SAM-SOM model for a hypothetical small-cell lung cancer diagnostics device 71

Figure 3.6 Illustrated Porter's model 78

Figure 3.7 Illustrated SLEPT model 81

Figure 3.8 SWOT model framework 86

Figure 3.9 From SWOT to TOWS 89

Figure 4.1 Marketing strategies based on the product/market matrix 103

Figure 4.2 Porter's four marketing strategies 110

Figure 4.3 The Hook Model 120

Figure 4.4 The Bullseye Framework 130

Figure 6.1 Simulation of the break-even point in a hypothetical SME 159

List of Tables

Table 3.1 Checklist to Assess a Company's Current Capabilities 33

Table 3.2 Major Steps toward Commercialization in Therapeutics—Scientific Track 35

Table 3.3 Major Steps toward Commercialization in Therapeutics—Marketing Track 36

Table 3.4 Sample Competitive Profile Checklist 63

Table 3.5 Demo of a Weighted Score Evaluation of Competitive Space 65

Table 3.6 Sample Serviceable Available Market Development 71

Table 3.7 Customer Funnel Table Sample 76

Table 4.1 Examples of Differentiation Tactics 112

Table 4.2 Summary of Marketing Strategies 113

Table 4.3 Example of an ICE Score Table 119

Table 4.4 The 19 Communication Channels Used in the Bullseye Framework 131

Table 4.5 Summary of Distribution Strategies 137

Table 4.6 Application of the AARRR! Framework 141

Table 5.1 Sample Implementation of a Marketing Strategy 148

Table 6.1 Application of the Advertising-to-Sales Ratio 161

Table 6.2 Marketing Metrics Cheat Sheet (If You Are Generating Sales) 163

Table 7.1 Resources for Advertising Guidelines Directly to Consumers (B2C) 173

Preface

> The essence of strategy is choosing what not to do.
>
> **Porter**

Welcome to the second book in a series on marketing in life sciences.

In this volume, we will be going over the basics of marketing strategy. While the first tome presented different techniques on how to gather and analyze data, it was more akin to a marketing toolbox. This book goes a step further, as the objective is to share perspectives and frameworks, from classics to cutting-edge ideas relative to marketing strategy for life sciences.

For many companies in life sciences, marketing does not take an important role in their business strategy. For some, it is not part of their business model, as it is not their intention to interact with consumers, expecting the company to be acquired long before the first client purchases its first product. For others, the business model is based on finding a distributor who will take most of the responsibilities for commercialization activities.

But developing your marketing strategy is essential to how you build your product. As Paulina Hill, principal at Polaris Partners, mentioned during her interview, "Sometimes the product will need to undergo a shift because the underlying technology is not optimal for the market the entrepreneur is targeting. Sometimes you may be able to keep your technology as is, but other times you may need to completely revamp the product to see it succeed."

This book provides several guidelines and inputs to help young entrepreneurs prepare their early marketing strategy. It is intended for

- Start-ups or small companies looking to develop, implement, and control their marketing strategy.
- Established marketing practitioners and other professionals from outside life sciences, looking to gain a basic understanding of how to apply marketing strategy essentials to life sciences.

To complete this book, we interviewed life sciences venture capitalists to get their perspective on marketing strategy. As such, I want to personally thank everyone who agreed to participate and discuss the topic of my book, and for their insight. Paulina Hill, principal at Polaris Partners; Lidija Marusic, investment manager at Innovacorp; Marc Rivière, general partner and CMO at TVM Capital; and Katherine Parra Moreno, VP MedTech, Digital Health, Healthcare and Venture Capital at Epic Capital Management. I express my deep gratitude for all your insights and giving the book an additional dimension based on your professional experiences.

I also want to thank my clients who I have had the great pleasure of working with and who graciously agreed to share information on some of the projects we did together throughout the years. A huge thank you goes to

- Sophie Chabot, CEO, JustBIO
- Tom Cohen, CEO, Nanopore Diagnostics, LLC
- Claude Leduc, CEO of MRM Proteomics
- Perry Niro, CEO of Pharmed Canada

Finally, I wish to express a genuine thank you to my wife, Corinne, and my children, Jaz and Chad, who encouraged me during the long weekends and evenings, helping me when they could. Their positive encouragement was appreciated and essential in completing this volume.

It is my hope that this book can help you on your journey to innovation.

Introduction to Marketing Strategy

A Strategic Introduction to the Importance of Marketing Strategy

We start our journey in marketing strategy by presenting the story of Sedasys, a revolutionary medical device developed by Johnson & Johnson. Sedasys is a computerized sedation and anesthesia system which observers believed would revolutionize operating room procedures by automating the anesthesia procedure during routine colonoscopies. This would eliminate the need for the presence of an anesthesiologist during some operating procedures. Having obtained FDA approval, Sedasys was promoted as a device which would generate savings to hospitals, patients, and insurers by replacing anesthesiologists in the operating room.

The device's mission did not go unnoticed by health-care practitioners, especially by those the device was designed to replace. By promoted Sedasys as a device that could replace medical specialists, promoters immediately raised the ire of key opinion leaders, who used their influence to plant the seed of doubt about the machine's efficacy and the dangers of automating such a delicate operation.

At that time, there were no reported technical issues with the device. But the device certainly did face an uphill battle for adoption in the hospital sector. Intense lobbying by stakeholders meant that Johnson & Johnson had to change its marketing position from "replacing" to "assisting" anesthesiologists in operating rooms and propose that it should only be used during simple operations, with an anesthesiologist present.

Some believe the device failed to gain traction in the market because the marketing message did not correctly address all the stakeholders. While hospital administrators loved the device (it was generating considerable savings, as the cost per operation was $150–$200 for using the device versus $2,000 for an anesthesiologist), anesthesiologists were directly threatened by the device. Many claimed that a machine could never replicate a human's care and diligence, an argument as old as the story of John Henry.*

So, what could have been done differently to ensure proper integration? Perhaps identifying anesthesiologists as stakeholders in the process (rather than defining them as a cost that hospitals could cut) could have helped the company identify the opportunities to collaborate. But hindsight is 20/20, and the goal here is not to criticize what happened, but rather use it to illustrate how careful market planning is invaluable to reaching and penetrating the market.

Marketing, Marketing Strategy, and Marketing Management

Before moving forward, let us take a moment to differentiate three key terminologies: marketing (in general), marketing strategy, and marketing management. The three terms

---

* For those unfamiliar with the story of John Henry, it is the folktale of a railway track worker, who, confronted by a machine which would replace him and his colleagues, challenged the company in a race to see who would be quicker, man or machine. He won, but his heart gave out. It is often used as a symbol in labor movements, as illustrative of the man versus machine struggle workers are faced with in our modernizing world. While the story itself is subject to interpretation (and its veracity is still questioned), the message remains quite colorful and powerful.

are sometimes used interchangeably, so taking a moment to understand each is essential.

Marketing is the most encompassing term as it includes the global business marketing activity. The American Marketing Association has defined marketing as "the activity, set of institutions, and processes for creating, communicating, delivering, and exchanging offerings that have value for customers, clients, partners, and society at large" [1]. Marketing includes a wide range of activities such as market research and advertising, and marketing departments often collaborate with other departments such as logistics, product innovation, and development as well as public relations to help the company reach its business goals. In some situations, marketing departments take the lead in building customer relationships, as customer engagement is paramount for creating retention and generating repeat purchases. Overall, marketing can be defined as all the activities that an organization engages in to attract, acquire, and retain customers.

Marketing strategy is defined as the ensemble of activities an organization performs to position its products and services (or even itself) in the market, so it can deliver value and achieve its business goals. Kotler adds that the marketing strategy lays out the target markets and the value proposition that will be offered based on the analysis of the best business opportunities [2]. As such, building a marketing strategy includes activities such as data collection, analysis, and planning. The goal of the marketing strategy is (1) to identify and articulate a competitive advantage and (2) to ensure and increase sales.

Marketing management encompasses the "process of planning and executing the conception, pricing, promotion and distribution of good, services and ideas to create exchange with target groups that satisfy customer and organizational objectives" [3]. According to Philip Kotler, "Marketing management is the analysis, planning, implementation and control of programmes designed to bring about desired exchanges with target markets for the purpose of achieving organisational objectives" [3]. This means that marketing management focuses on the **practical application of the marketing strategy** and on the management of a company's marketing resources and activities. Effective marketing management will use a company's resources to increase its customer base, increase sales, improve customer's opinions of the products and services, and increase the company's perceived value.

In life sciences, marketing strategy and marketing management are not always a primary concern. For example, not every company has a business model which includes sales to end users as the corporate objective. Some companies plan to out-license or sell their product long before they get to market. For others, sales will be limited to a single buyer (i.e., government agency) or will be the subject of a lengthy procurement process (i.e., hospitals and purchasing groups). The traditional consumer is not always the objective for companies in life sciences.

But these challenges should not exclude the development of a marketing strategy. If a company wishes to develop partnerships, it will need to demonstrate that its products or services are answering a specific market need, that the company has put some thought into how the customers will be reached, that a need has been identified, and that the company understands how to reach its market. As such, entrepreneurs will find that many conversations with potential partners and investors shift to the conversation around the marketing strategy and commercialization of their technology.

The Importance of Marketing Strategy

A sad truth in commercialization is that consumers do not always buy the best products. Consumers are sometimes irrational, sometimes biased, and sometimes just plain lazy. Many factors come into play during the purchasing process that extend well beyond the product's performance and influence the consumers' behavior. These factors include attributes such as price, advertising, word of mouth, product placement, color of the product, convenience, nostalgia, celebrity endorsement, packaging, availability, and logo.

One of the most important roadblocks a company might face is consumer inertia. Consumers will make the same purchases, week after week, month after month, not really reflecting on their purchasing process until there is a sufficient change that warrants reflecting on their purchasing decision. A price saving of 5% might not change most consumers' purchasing patterns, but a product that works twice as fast might.

As such, while a new innovative product might "technically" and "rationally" be better than the existing products on the market, it could take considerable effort to shift consumers from their existing purchasing patterns. I have met more than one entrepreneur who had an innovative product which, while technically superior to the existing "gold standard," was not enough of an improvement to warrant consumer interest. More than 40% of products fail once they reach the market, even if they have been successful all the way through test marketing [4]. That means that even though companies have rigorously researched their market and tested their concepts, some elements of commercialization failed.

As such, establishing a solid marketing strategy is essential. Be it targeting individual consumers or large organizations (such as a hospital or a government agency), the marketing strategy will be a critical tool to help the entrepreneur reflect, strategize, and prepare the march to market. Ultimately, a marketing strategy will allow the entrepreneur to

- Enhance the understanding of the market and identify value opportunities.
- Enable strategic choices and optimize innovation by selecting the best market niches and segments to target.
- Create a value-based strategy by identifying a customer value proposition for its selected markets relative to competitors.

- Develop a customer value strategy and positioning, establishing effective strategies relative to customers, collaborators, and competitors.

> **On the importance of preparing a marketing strategy**
>
> Understanding reimbursement policies, how to position the product, how to demonstrate and share evidence that the innovation either costs less, or provides more benefits, or both, brings a lot of value to the conversation. While we don't always expect a fully detailed reimbursement strategy and commercialization plan, it is important that the entrepreneur shows that he/she understands the full life cycle of the product. It's not just about how it works, but also how it is better from what already exists, and how it will get in the hands of people who need it.
>
> *Lidija Marusic, Investment Manager, Innovacorp*

Business Strategy versus Marketing Strategy

It is also important to understand the difference between a company's business strategy and its marketing strategy, as there is a lot of potential overlap between the two. Basically, a marketing strategy is a subset of a company's business strategy. An organization's business strategy includes a more global perspective of an organization, while the marketing strategy is focused exclusively on the marketing function and will both be impacted by and impact on the global business strategy.

Essentially, the business strategy is primarily concerned with the profitability of the organization. As such, the business strategy is articulated around topics such as cost controls, financing, and generating revenues. All of these fall in the purview of the business strategy. Since there are more often than not revenue considerations, the business strategy will often significantly overlap with the marketing strategy, which is primarily constituted around revenue-generating activities and interacting with the customer.

Of course, both strategies are synergistic and constantly impact one another. While a company's business strategy can impact a marketing strategy, the reverse is also true. As such, if a marketing strategy is effective and is generating more benefits than other business units, a business strategy could be impacted to the point that resources are reallocated internally. For example, if a marketing strategy has increased sales significantly, it could impact previous production and distribution decisions. Conversely, if the company is making several decisions around the product in its business strategy (e.g., reducing the quality of components used in manufacturing the product), many of the marketing decisions in the marketing strategy will be impacted (e.g., a change in product quality would directly impact a branding message of a "superior quality product").

Furthermore, a successful marketing strategy will have a direct impact on an organization's business strategy and other business functions. As the marketing strategy is articulated by using a plethora of specific marketing tools (such as branding, advertising, public relations, and customer relationship), these will directly impact a host of other functions in the organization. For example, a company with a successful image and brand will have more success in hiring and retaining key resources or securing deals with key suppliers.

As such, the marketing strategy is a key component in a successful business strategy, and a key element of a successful company is maintaining an alignment between the two.

References

1. Grönroos, Christian. 1989. Defining Marketing: A Market-Oriented Approach. *European Journal of Marketing* 23(1): 52–60.

2. Kotler, Philip, & Keller, Kevin Lane. 2015. *Marketing Management,* 15th edition, Pearson Higher Education, Englewood Cliffs, NJ.
3. Kotler, Philip. 1997. *Marketing Management: Analysis, Planning, Implementation, and Control*, Prentice Hall, Englewood Cliffs, NJ.
4. Hustad, Thomas. 2013. *PDMA History, Publications and Developing a Future Research Agenda: Appreciating Our PDMA Accomplishments—Celebrating People, Lasting Friendships and Our Collective Accomplishments.* Xlibris Corporation, Bloomington, IN.

Author

Jean-Francois Denault has been working with innovators and entrepreneurs in life sciences as a professional consultant for over 15 years. Through the years, he has worked with more than 50 different clients in life sciences (including larger companies such as Johnson & Johnson, Procter & Gamble, AbbVie, and Denka Seiken). His clients are located all over the world, having completed projects with clients in more than 25 different countries.

Denault specializes in life sciences. As such, he has completed projects related to pharmaceuticals, biotechnology, medical devices, nutraceuticals, and health care. Most of his projects have been in the market research, marketing strategy, and competitive intelligence space. He possesses a graduate degree in management consulting and an executive MBA as well as a graduate degree in organizational communication.

Besides having authored this present volume, Denault has previously written *The Handbook of Market Research for Life Science Companies*. He is a member of the editorial boards of the *Journal of Brand Strategy* and the *Journal of Digital & Social Media Marketing*, and has written a dozen articles for various publications. He is on the advisory boards of several start-ups (including TOWWERS, Marshall Hydrothermal, and JustBIO) as well as a member of Pharmed Canada's CMO-CDMO Strategic Advisory Committee. Finally, he is an active member of his community and has been the president of the Lanaudière Alzheimer Society since 2012.

Denault is based in Montreal, Canada, and can be reached at jf.denault@impacts.ca.

Chapter 1

Marketing Strategy Road Map

1.1 Planning toward Strategy

Many entrepreneurs want to develop a marketing strategy without first looking at their surroundings. More than once, I have been approached by a CEO of a start-up who wanted me to prepare a marketing strategy for his organization, but he was unable to answer some basic questions: What are your organizational objectives? What are your competitors doing? Who is your client? In some situations, even "What is your product?" was problematic to answer.

As such, I have found that there is a tendency to develop a marketing strategy without context, and then adapt, modify, and change the strategy as the company grows. But if you are changing your strategy each time new information becomes available, you are not necessarily being nimble and adaptive, but might be overly opportunistic and reactive. Your marketing strategic plan should be prepared around short- and long-term goals, containing tactics that are adapted after a careful examination of control values.

> **On the importance of marketing strategy**
>
> Marketing strategy is very important. It can be very complex, especially if you are going to sell a "low margin/high volume" product. If you have designed a "high margin/low volume" product (like a hospital software solution), then your go-to market plan might not be as complicated: you identify your 20 key potential clients, and then develop the relationships accordingly. But if you are looking at a mass market situation, the distribution strategy, the commercialization plan, and the product positioning are all going to be very important.
>
> *Katherine Parra Moreno, Vice President of Business Development, Epic Capital Management*

1.2 Planning Your Marketing Strategy

In this book, we have simplified the preparation of the marketing strategy into four simple steps: complete your **situation analysis** (Chapter 3), **develop your marketing strategy** (Chapter 4), prepare **implementation and control** (Chapter 5), and set **marketing metrics** to evaluate the progress (Chapter 6). This is preceded by a short chapter on **market research** (Chapter 2) and is followed by a chapter discussing unique perspectives (marketing in life sciences, marketing health-care services, and marketing health-care digital products) (Chapter 7) (Figure 1.1).

1.2.1 Performing Market Research

Before preparing your marketing strategy, it is important to have the information you need to make the right decisions. Market research will be crucial, and it will be an ongoing process throughout your marketing strategy effort. As such,

Marketing Strategy Road Map ■ 3

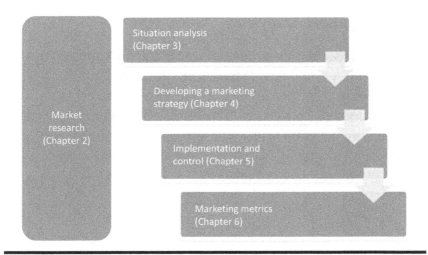

Figure 1.1 Basics of the market research process.

in Chapter 2, we will be going over the basics of market research, by exploring two aspects.

1.2.1.1 Planning Your Data Collection

Building your data collection tools correctly is crucial to ensure data is collected in a consistent manner, especially if multiple people are collecting the data independently. As such, I propose the use of four simple steps to building your data collection tool which are as follows: (1) define the information required, the target respondents, and the data collection tool you will use; (2) build your question bank; (3) triage and select questions; and (4) test the data collection tool prior to use at large.

1.2.1.2 Overview of Market Research Tools

There are many ways to collect your data. To cover the basics, we will be going over some of the more popular ones in life sciences which are in-depth interviews, online surveys, and focus groups. A brief discussion on secondary data collection techniques will conclude this chapter.

1.2.2 Situation Analysis

Your situation analysis is built using your information, which can be both internal and external to your organization. We will be going over the different types of information that you will need to collect and how to classify them so you can have a more accurate understanding of your company and its business environment. The main content of Chapter 3 splits into internal and external analyses.

1.2.2.1 Internal Analysis

Internal analysis is performed by examining the organization, so it can determine its own strengths and weaknesses. Some of the different things we will examine include processes, capabilities, and resources, as well as review the company's mission and vision. Also, the business model is of explicit interest. Are you an integrated company or are you using a research model? Are you licensing your technology or are you going to market yourself?

1.2.2.2 External Analysis

External analysis is the analysis of factors that are outside the organization, but directly impact its functioning. While the company (as an individual organization) usually has very little control over these factors, it is very important to understand them. We will spend some time focusing on the key question, "Who is your customer?" so we can really expose the dynamics where the end user of the product can be different than the person who will be paying for it, and discuss the potential disconnect between the person who benefits from the product and service, and the payer. We will also go over the different actors' interactions (and their specific interest) with life science products: the patient, the hospital, the primary caregiver, the private insurer, the group purchasing organization, and the government agency. Following this, we will take an in-depth look at competitive analysis, how to build a relevant market assessment and how to prepare a useful environmental assessment, which includes both the microenvironment (Porter's Five Forces) and the macro-environment (the Social, Legal, Economic, Political, and Technology [SLEPT] model).

1.2.3 Developing a Marketing Strategy

Developing a marketing strategy is an extensive venture, so we will be dividing it in multiple sections. The first section will deal with the topics relevant to your marketing strategy vision, followed by a short discussion on traditional company marketing models, and identify which model is the most relevant for you. After these two sections, the next two sections will delve into the more practical applications of marketing strategy and describe the fundamentals such as the pricing strategy, distribution strategy, and promotional strategy. Finally, due to its importance, we conclude with a distinct section on digital marketing.

1.2.4 Implementation and Control Mechanisms

The implementation of your marketing strategy is the series of steps necessary to ensure you achieve your goals. It is a practical approach, the real-world application of your marketing plan. It is also a realistic assessment of how you will deal with roadblocks, as they emerge.

Meanwhile, control deals with the tools you need to measure the effectiveness of your strategies. As such, we will go over different control tools that you can plan to implement to ensure the resources invested in marketing and sale are properly generating revenues. This is split into two chapters. Chapter 5 deals with the more practical aspects of implementation and control mechanisms, whereas Chapter 6 focuses on the marketing metrics themselves for marketing decision making.

1.2.5 Final Notes

Chapter 7 addresses specific situations such as the marketing in life sciences, the marketing of health-care services, and the marketing of health-care digital products. For example, when discussing marketing strategy for services, we must acknowledge that there is a distinct approach for selling services versus products; hence, it is worth presenting the differences and similitude between traditional products (drugs, devices) and newer service technologies. The sale of apps and newer technologies also creates distinct challenges that should be accounted for.

Chapter 8 wraps up our journey toward marketing strategy while addressing the topics that will shape future marketing endeavors in life sciences, discussing the disruption of traditional health-care systems by patients.

Chapter 2

Overview of Market Research

The situation analysis is based on the information you have access to. The more precise the information, the better the decision making. When you start the process with some information on hand, it will most likely be necessary for you to engage into some market research activities to collect the missing data.

Market research is done for two audiences. First, it is done for the organization itself so it can make optimal decisions, e.g., when the lack of information could cost the organization more compared to the cost to acquire the information. Second, market research is done to convince potential partners and investors, by demonstrating the validity of an organization's marketing strategy. This information will be necessary when making decisions and planning strategy.

In the text that follows, we will be reviewing some basic market research concepts, followed by the basics of the most popular market research techniques (both primary and secondary research). More information on how to plan and execute your market research, as well as analyze the data you gather, is available in the other book of this series, *The Handbook of Market Research for Life Sciences* [1].

2.1 Basic Market Research Concepts

Market research is full of dichotomies. To fully understand the market research process, we will be going over some of the most important concepts, which are primary versus secondary research, quantitative versus qualitative research, and inch-deep/miles-wide versus deep dive research projects.

2.1.1 Primary and Secondary Market Research

Primary market research is a market research activity in which the entrepreneur is actively engaged in doing research and creating data. Simply put, it is the collection of data that did not exist until the researcher completed the market research activity. Some of the most popular tools are web surveys, in-depth interviews, or focus groups. Although more costly than secondary data, it is tailored specifically to the researcher's needs and will belong solely to him and his organization (meaning competitors will not have access to this specific information).

Secondary market research is the collation of data that already exists. For example, it could be collected through a web search or by aggregating news posts or blogs. The researcher collects and then transforms the data into something coherent and useful. Although less costly, it is not always tailored to the researcher's needs.

2.1.2 Quantitative and Qualitative Data

Data can be quantitative or qualitative in nature. *Quantitative data* refers to data that can be measured and numbered. "Counting" the number of potential clients for a product, calculating the number of products or doses of a drug a consumer uses each day, or the average distance a patient is willing to travel to visit a specialized clinic are all different types of quantitative data. People in life sciences are usually quite familiar with quantitative data to quantify the technical aspects of products. Quantitative data in market research is often used to determine the market size and identify market segments and opportunities. Some data collection tools are better adapted for quantitative data. For example, surveys (both online and in person) are usually the best ways to generate an important quantity of quantitative data.

Meanwhile, *qualitative data* is data that is subjective and subject to interpretation. It can include anything from stories to discussions, observations, or pictures. Some examples of qualitative data include personal reasons for preferences in consumer products, the impact of quality on customer purchasing patterns, or the impact of packaging color on the purchasing decision. Data collected through interviews, focus groups, and observation is usually of qualitative nature. In general, quantitative data is perceived as "more real" and is easier to convey to an audience, whereas qualitative data is very useful to understand and contextualize the story behind the numbers.

To illustrate this, let me share a project I did a few years ago. My client, an Australian ad agency, was working on a promotional campaign for a big pharmaceutical company. To prepare the right promotional message, we had surveyed over 3,000 consumers on their usage of pain killers. The quantitative data demonstrated what the preferred brands were, but without context, we could not understand why the number one preferred brand was the one my client perceived as cheapest and less efficient. It is only after analyzing the qualitative data in the survey relative to why consumers made these purchasing decisions that we were able to uncover patterns in decision making (the major categories of reasons people gave were centered on topics such as the family's choice, health reasons, routine, price/sales, and advertising). To deepen our understanding, we monitored spontaneous online discussions relative to the brands. We then found passion around the preferred product due to fewer secondary effects, debates on the home brands versus branded products, and patterns on how consumers perceived competing brands. We found that the brand that was preferred by the consumers, while perceived as less effective, was recommended more often, whereas the second most recommended brand was perceived as being tougher on pain, yet harder on digestion.

2.1.3 Miles-Wide versus Deep Dive Research

When approaching a market research project, a researcher will have to decide if he is going for a miles-wide approach or a "deep dive."

The miles-wide inch-deep approach indicates an overview of a segment, an industry, or a competitive landscape. As such, the research project is deployed in a way to gather information on as many multiple data points as possible, simultaneously. By its nature, it is very exploratory. By the end of the market research initiative, you will possess a high-level overview of the specific topic.

The deep dive approach focuses exclusively on a specific predefined topic. When doing this type of project, the top trends, competitors, or issues have already been identified, and the researcher deploys his energy to researching each of these topics in a very specialized approach. This means that rather than interviewing generalists, you will be interviewing key opinion leaders and consulting specialized resources on a specific theme.

Many projects will be a convergence of these two types of research, starting with the inch-deep/miles-wide and concluding with a deep dive on the most interesting targets. Remember that it is important to align expectations with available resources.

2.2 Preparing Your Market Research Plan

Coherent and valuable market research follows a systemized approach. As such, like many other processes in life, it all starts with building a plan. While it is tempting to jump directly into "market research" and start collecting data, a detailed market research plan ensures that the data collected will be consistent and useful. As we will see, market research must be planned beforehand to ensure consistency in the data that is collected, as well as formalize the end point. Preparing your market research plan breaks down into two distinct steps: identifying and formulating the problem, and then determining the research design.

The first step of the market research process is to identify and formulate the problem (or the opportunity). By formally defining the problem, the market researcher will focus his research effectively, ensuring all participants share the same vision and objectives for the project. As such, the problem identification step will usually involve discussions with decision makers, a review secondary data, and conversations with key opinion leaders.

The topic of research is usually defined in a few words. For example, it could be to identify emerging market opportunities for a new technology, and the size and segment of the current market, or to develop a customer profile (around their specific needs, issues, and problems).

The next step is to determine the research design. It is the approach that you will use to collect your data and guide you in choosing the specific methods you will use to collect the information you need. Some key questions you will answer at this step are as follows:

- Which method(s) will I use to collect data?
- How will I connect with my data sample? Who will I need to connect with? How can I connect with them? Will I need to incentivize them? How?

- Which data collection tools will I use (telephone, in person, Internet)?
- What is my total budget (both monetary and timewise)?

The data collection phase splits into distinct phases. First, there is a design step in which you will design your sampling plan and tools, followed by the collection of data.

The sampling plan is the detailed framework of who will be contacted and what the expected sample size is. The sample size is crucial for the validity of the information you collect.

Once the research question has been designed, and the methodology decided upon, it is time to design the research tool. For example, if you have decided to do interviews, you will have to design an interview guide. Building an interview guide ensures consistency between interviews and between interviewers, and it is also a useful tool to remember topics during the interviews themselves.

Ideally, you should test your research instrument before using it at large. Test your interview guide with a few potential participants: you might find that some questions are redundant, some questions are missing, and some questions are misunderstood by your target audience. It is much more cost-effective to find this out at this stage than at the data analysis stage.

Finally, you are ready to start collecting data, which is often the most time-consuming step of the market research process. Once you have reached the end point of your data collection, it is time to start analyzing data and use the frameworks.

2.3 Collecting Data—Primary Research

Primary data is information that is generated directly by the market researcher to answer his research question. For example, when he is doing an interview, managing an online survey, or performing client observations, he is gathering primary data.

Generally, primary research is costlier to generate (in terms of both time and resources), but it is customized for the researcher's needs. If he has correctly designed his tools, he should be able to solve his research problems. Also, the data he collects is proprietary, so it belongs to the organization exclusively, becoming a competitive advantage.

2.3.1 Data Collection Methods

In this section, we will be going over the basics of the three main market research data collection tools. These overviews are useful to understand and select the instrument that you need for your project. The methods we will be reviewing are in-depth interviews, focus groups, and online surveys.

2.3.1.1 In-Depth Interviews

In-depth interviews are interviews that are done one-on-one between the researcher and the participant. These interviews consist of mostly open-ended questions. The objective is to explore the topic in a semiformal format, gathering qualitative information. While in-depth interviews are costlier (in terms of time and money), they present a number of advantages over questionnaires and online surveys. For starters, interviews give more opportunities for the researcher to motivate the respondent to participate in a truthful manner, and to not abandon the interview halfway through. Also, interviews allow more flexibility in exploring secondary topics as opportunities in the interview emerge. The more exploratory the topic, the more useful the in-depth interview is, as it allows the researcher to change the order in questions, to prioritize some topics if time is short, or to go more in depth if the participant is resolved to have a rich narrative on a specialized topic.

There are a number of actions a researcher can do to enhance the interview process. First, *prepare your interview*: conduct a quick due diligence on the interview target prior to your interview to identify potential specialization and fields of specific interest. Also, *record the interview*: when you record the interview, you will be able to focus on the answers the participant is providing, as well as ask questions and explore topics with your interviewee, rather than spend time taking notes. Finally, *listen to your interviewee*: you are trying to collect data, which means the participant needs the opportunity to share his information. Be careful not to spoon-feed the answers that you are looking for. This leads to bad data collection, and ultimately does not reflect real market conditions.

In-depth interviews in life sciences are a popular way to get the information you need, especially when dealing with topics which are sensitive in nature. It is a great way to speak to people confidentially and get their views on health-care topics, such as their personal health and those of loved ones, and their use of pharmaceuticals. Interviewing doctors and medical personnel can be especially challenging as these individuals are very solicited for their time. As such, you might have to set aside an important per diem to get them to participate.

2.3.1.2 Focus Groups

A focus group is a small group of individuals brought together to discuss a specific topic. The added value of a focus group (versus individual interviews) is that the interaction between individuals generates a wealth of information for the researcher. As such, focus groups are useful to gain a better understanding of what people are thinking and why they are thinking about it. It is also interesting as many participants will generate more information in a group setting where they feel safe and they do not feel they are the sole focus of the interview. Finally, the information gathered in a focus group can be very useful to design a follow-up quantitative questionnaire, or to interpret the information gathered from a quantitative research project.

The main issue with focus groups is similar to issues you find in peer groups. As such, social pressure (the desire to conform to the group and not disagree), an individual's domination of a group, or the halo effect generated by key opinion leaders can all impact the quality of the interactions and the information generated. It is the role of the moderator to step in, to rebalance the focus group, and to ensure that it does not become biased. The other issue to remember is that while focus groups can be used to evaluate a group's feelings or views on a topic, it cannot be used as a final decision tool. It is exploratory in nature, not statistically valid, and the information you gather, while invaluable in interpreting the existing data or setting up more research, cannot be an endgame in itself.

Ideally, a focus group will have eight to ten participants, as it needs to be sufficiently large to generate dynamic conversations, but not too large as to leave some participants out or become difficult to manage for the moderator (as parallel discussions start to emerge). About 45–75 minutes is the ideal time frame for a focus group: if the time is too short, you will not generate any deep insight; if it is too long, you risk participant fatigue and the participants quickly agree with one another in the hope that the focus group will end. Three to four focus groups are usually necessary to fully explore a topic; after that, you may find the same information being repeated.

In life sciences, focus groups are especially useful to interact with end users (doctors, nurses, laboratory technicians) as well as get feedback from patients. As such, they can generate valuable information relative to marketing, branding, competitors, and product issues. Nonetheless, some researchers have found that focus groups are not an ideal environment for eliciting emotional information from physicians, because the image doctor's want to project to others is typically one of a rational decision maker [2]. Also, it can be especially challenging to reach and recruit specific participants.

2.3.1.3 Online Surveys

The rapid pace and development of technology has created new opportunities for collecting data. As such, the use of online surveys has grown immensely in popularity. They are cost-effective and simple to use; if done properly, they can reach a wide range of the population, allowing participants to complete the survey quickly on their own time with little effort. Most online surveys today are done using a web-based survey tool.

There are considerable challenges in getting participants for a life sciences survey, particularly clinicians. A study conducted in 2015 found that an online web survey targeting clinicians got a 35% participation rate, with deep variances across specialties, ranging from 46.6% (neurology/neurosurgery), 42.9% (internal medicine), 29.6% (general surgery), 29.2% (pediatrics), to 27.1% (psychiatry). Lack of time and survey burden were the most common reasons for not participating [3].

Another study found that general practitioner survey rates could be increased with incentives (larger and upfront, if possible), peers precontacting targets by phone, personalized packages, and sending surveys on Fridays [4].

There are several things to remember when building a web survey to increase your response rate. First, *keep the survey short and simple.* This helps to reduce user attrition. Second, *be straightforward about the time to answer.* Announce upfront the length of the survey and, if possible, use a progress bar on top of the survey to keep participants engaged. Finally, *optimize your survey for mobile devices.* Use a survey platform that will optimize your survey for mobile platforms: more people use mobile devices to do mundane tasks while waiting, and if your survey does not properly display on a mobile device, they might simply drop out and move on.

2.4 Secondary Research

Secondary research is the collection and collation of information that is published and publicly available. Also called desk research, it is frequently done to explore a topic before engaging in more expensive primary research, or to quickly gain a summary understanding of a topic without engaging too many resources.

The advantages of doing secondary research (versus primary research) are numerous. First, secondary research is considerably less expensive than primary research. Also, it is considerably faster to complete as it does not depend on third parties (such as recruited participants, organized focus groups, or enrolled survey participants) to obtain the information, and the sample size of third-party research reports will often be quite considerable. Furthermore, due to scope and reputation, third-party information often has more perceived authority and impartiality than "in-house" research. An assessment such as "BCC Research estimates that the global advanced drug delivery market should grow from roughly $178.8 billion in 2015 to nearly $227.3 billion by 2020, with a compound annual growth rate (CAGR) of 4.9%" [5] has more credibility than most in-house estimates. Finally, as mentioned earlier, secondary research is very useful to orient and define primary research; a researcher will often start a market research project by doing a quick market review to identify some of the main trends and concerns before diving into primary research.

There are some disadvantages to secondary research. First, the information is not always personalized to an organization's requirements, and it is quite difficult to find data for emerging fields. Reports on nanomedicine are plentiful, but developing a specific application merging nanomedicine and information technology means that secondary research will be quite scarce, and that the organization will need to either (a) extrapolate from generalist research or (b) conduct primary research. Also, the data might be outdated, limiting its usefulness.

Finally, most of the time, the original data used in secondary data is unverifiable; it is quite difficult to spot errors in data collection or dispute the way data was analyzed in a consolidated report.

There are two types of secondary data: external and internal secondary data. External secondary data is information gathered from outside the organization. This includes anything from government statistics to media sources. Internal secondary data is data that the organization is generating itself. It could be data collected from customers' feedback, accounting and sales records, or employee experiences. While it is possible to do secondary research using non-Internet sources, the bulk of our suggestions are related to this medium for the ease of use, convenience as well as cost.

Finally, there is a distinction between active and passive secondary research. Active secondary research takes place when the researcher is actively searching for information, whereas passive secondary research is the use of tools and software to automate data collection.

2.4.1 Active Secondary Research

Active secondary research takes place when the market researcher is dynamically searching for information. In the text that follows, we will be going over the most popular and pertinent sources of data available. All the suggested sources are web based.

While there are some ethical considerations for what information the researcher can use, a simple rule of thumb is that any information made available to the public is fair game for collection and review. If the method used to obtain the information is not commonly available to the public (such as using a former employee's password to access a restricted website area), then it is not only unethical, but most definitely illegal.

2.4.1.1 Popular Sources of Data Online

- **Government data**: Government agencies generate large bodies of information which can be used by researchers. Most of this information is free to use and can be useful at the start of a research project. Government data is usually statistical in nature and is very useful when building marketing models or trying to understand the nature of a market. Some key databases include the National Center for Health Statistics (www.cdc.gov/nchs/hus.htm), the World Bank Open Data (data.worldbank.org), and Eurostat (European Union—ec.europa.eu/eurostat).
- **Public company data**: Companies publish a lot of useful information online. Often, companies will do market research and will publish results as some consolidated information in corporate documentation. As such, reviewing this publicly available data is another useful way to start your research effort. Some of the documents that can be reviewed online include annual reports, company pitch decks, press releases, and video product presentations.
- **Print media**: Print media sources are published on a regular schedule by specialized companies. These include magazines and newspapers, and include both their printed/physical format and their internet counterparts. There are a number of trade magazines that are published on a regular basis which are of use to market researchers. Most of them are free, and access to their archives is public most of the time, although some do monetize their archives. Popular ones include Fierce Biotech (www.fiercebiotech.com) and MD+DI (www.mddionline.com).
- **Social networks**: For the purpose of this book, "social networks" are defined as dedicated websites or applications where users with a common objective aggregate

and participate in discussions. These discussions create a network of social interactions, as users share messages, comments, information, opinions, experiences, and more. Social networks are a useful way to get the customer's pulse on a topic. Analyzing the comments found on social network pages allows the researcher to understand how people view a brand or their perception of a specific topic. It is also possible to use social networks to recruit participants for surveys and focus groups, as well as interact directly with them and engage in one-to-one conversations.

- **Trade and industry groups**: Trade and industry groups are organizations representing multiple firms (private companies, government agencies, universities, consultants) in a common commercial activity sector. Some of the larger associations produce (or sponsor) reports relevant to their industry, which can be of use during market research. One of the advantages of the reports produced by these organizations is that they are third-party reports and (relatively) unbiased, but the reports are not always freely available and are sometimes reserved for members. Some of the main trade groups you should be keeping an eye on (depending on your specific industry sector) include the Biotechnology Innovation Organization (www.bio.org), PhRMA (www.phrma.org), the American Hospital Association (www.aha.org), and the Medical Device Manufacturers Association (www.medicaldevices.org).

2.4.1.2 Using Search Engines to Look for Information

Searching through the Internet using a web research engine is usually the first step of secondary research. Here are a few tips and tricks to make the research effort more efficient:

- *Go beyond Google*: If you are not finding the information you need, you can try using another search engine to obtain different search results. Some of the interesting alternative search engines include the following:
  - Bing (www.bing.com) (which is reported to have a better video search option).
  - Board Reader (www.boardreader.com) (which specializes in the user point of view by searching through forums, message boards, and Reddit).
  - Slide Share (www.slideshare.net) (a cornucopia of PowerPoint presentations, slide decks, and webinars from past conferences).
- *Look for corporate web DNA*: I originally found this technique referenced by Leonard Fuld in *The Secret Language of Competitive Intelligence.* It is based on the concept that every organization develops its own brand of corporate speak or pattern. It is akin to corporate web DNA. He defines it as "a unique pattern of words and phrases that form the substance of a company's website, its press releases and its advertisements." [6] As such, if the researcher can identify a group of unique words or jargon as potential corporate DNA, he can then proceed to research the web using the abovementioned terminology, grouped between two sets of quotations. As an example, using Medtronic's "Transforming technology to change lives" slogan to search the web brings up a series of white papers (old and new), job offers (both current and expired), as well as customer testimonials.
- *Look to the past*: Sometimes, a market researcher will be looking for something specific, but will conclude that the information is no longer available online. For example, it might be an old press release that a competitor has pulled from his website, information on a previous partnership that has been quietly ended, or specifications

on discontinued products. In these cases, the website Archive.org (an Internet Archive which is a nonprofit digital library offering free universal access to all) is especially useful. Archived in their public database are historical web snapshots of the company's website, which can include pages, attachments, and more. While a researcher might not have access to each version of a company's website, there are often several snapshots taken throughout the year, enabling the researcher to identify key information that has been removed online.

2.4.2 Passive Secondary Research

Automated Internet research tools are a boon to market researchers, as they automatically monitor and report on specific information topics. We will be going through some of the most interesting tools researchers can use to automate their market research.

- **Rich Site Summary (RSS) feeds**: RSS feeds are a simple method to aggregate data generated by specialized websites and efficiently supply researchers with up-to-date information on specific topics. The advantages of using RSS feeds for research and continued monitoring are multiple. First, RSS feeds save time; you can quickly subscribe to the feeds you are interested in and quickly scan aggregated data without having to visit every single website every time. Also, as RSS feeds update themselves automatically, you get information as it becomes available. Three types of RSS readers exist: web-based readers (which you access through your web browser) such as Feedly (www.feedly.com) and NetVibes (www.netvibes.com), client-based readers which you download and install on your computer such as RSSOwl (www.rssowl.org), and those that integrate into your web browser (Firefox and Internet Explorer both offer this option).

- **Google Alerts**: Google Alerts is a service offered by Google. It is useful to automatically monitor a topic by setting up a search alert. Once setup, the Google Alert sends search engine search results by email to researchers as they occur, or on a predetermined basis in the form of a digest (once a day or once a week), at a predetermined time. To create a Google Alert, the user only needs to go to the website www.google.com/alerts, type in his topic of interest, and customize the information feed requested (frequency, where the information will be collected from, the number of results wanted each period, and which email will be receiving the information). It is possible to edit an alert if needed or delete the alert if it is no longer needed.
- **Social media tracking**: There are a number of tools and search engines that are specifically designed to monitor social media. These tools search through popular social media websites such as Twitter, Facebook, and user-generated content such as blogs and comments to generate reports which can be used to identify underlying consumer trends. One of these popular tools in this space is Keyhole (http://keyhole.co/). This monitoring tool keeps an eye on keywords and hashtags across Twitter and Instagram and can be useful to quickly identify a popular public key opinion leader on a topic, which can then be engaged further for market research. It can also be used to identify geographic trends and estimate overall consumer sentiment. Another useful tool is Warble (https://warble.co/), which monitors Twitter and sends a daily email report directly to the chosen email account. It automates the process of monitoring Twitter, which is pretty important due to the important mass of content that is generated daily on it.

2.4.3 Internal Secondary Data

Internal secondary data exists and is stored inside the organization. It is available exclusively to the organization, and is usually generated and collected during normal business activities. Internal sources of data should always be investigated first because they are usually the quickest, most inexpensive, and most convenient source of information available.

Sources of internal secondary data include the following:

- *Sales data*: If the organization is commercializing its product, it has access to an invaluable resource, its own sales data. This data is usually collected by the organization and organized in a way that is useful and extractable by a market researcher. Some of the data the researcher can look at includes sales invoices, sales inquiries, quotations, returns, and sales forces business development sheets. From this information, territory trends, customer type, pricing and elasticity, packaging, and bundling impact can be inferred. This data can be useful to identify the most profitable customer groups and which ones to target in the future.
- *Financial data*: All functioning organizations have accounting and financial data. This can include cost of producing, storing, and distributing its products. It can also include data on research and development costs and burn rate, and can be used to calculate valuable ratios (see Chapter 6 for more information on marketing ratios, and on how to develop and interpret them).
- *Internal expertise*: Midsized organizations will often have inside expertise like personnel who have been with the organization for a while. These individuals can be tapped and interviewed to get more information on past initiatives, products, lost customers, or any other topic often referred to as the organizational memory; they often have a wealth of undocumented knowledge that can be

harvested. They might be aware of internally produced reports that might be of use, past projects, or failed product initiatives.

Some of the weak points of internal data are inaccuracy, either due to the fact that it might be dated or the way it was collected. Also, while most of the time data can be ported from internal systems to market research data analysis tools, some legacy system might make data conversion especially challenging. Finally, there are confidentiality issues: some companies employing third-party researchers might hesitate to "open the books." In this case, it could be possible to share either consolidated data or limited data sets.

2.5 A Few Words on Ethics and Market Research

We presented and discussed some of the tools a market researcher can use to collect data. But there are a few things to remember relative to ethics before engaging in market research.

First, it is very important to be honest when you collect data, *identifying yourself and describing why you are collecting the data.* Misrepresentation is a huge issue in data collection, and it is tempting to do so in an attempt to ease the data gathering. For example, some researchers will pretend to be a potentially interested client or pose as a student gathering information for a school project. This is clearly unethical. Instead, efforts should be spent identifying targets which have the information and are more likely to want to share it such as academics, technology vendors, advertising agencies, and journalists.

Also, *be neutral when asking questions.* It is very easy for a market researcher to influence the participant's response. Asking leading questions can cause a participant to answer in a specific way. Even agreeing with a participant rather than impartially acknowledging his answer can influence the participant. Remember that leading participants might get you the answer you want to hear, but will not necessarily reflect the real market's appreciation of your product. Wouldn't you rather find out the real market conditions for your product during the market research phase, rather than shaping market research to fit your preconceived ideas, and then failing during commercialization? *While market research shapes our vision of the market, it does not change the true nature of the market.*

Furthermore, *respect the confidentiality of the participants.* If you have given them the assurance that you will protect their responses, be ready to do so. If you believe you cannot assure the confidentiality of the participants, or if you do not intend to (by sharing raw data to other stakeholders), be upfront with participants so they can opt out.

Finally, *primary market research is not a commercialization activity.* Engaging participants in market research and then trying to sell a product midway during data collection undermines market research as a whole. In some cases, a client might express interest in a product you are researching. When this happens, my approach is to ask the participants, "You seem to have some interest in the product/service we are discussing. Would you like me to refer you directly to the company as an interested party? Do you accept that your coordinates be shared directly to the appropriate person?" As such, the participant is authorizing you to share his information and his interest. You are serving both parties and, with consent, are relieved of other obligations you might have (such as keeping the data anonymous).

References

1. Denault, Jean-Francois. 2017. *The Handbook of Market Research for Life Sciences*, 1st edition, Productivity Press, Boca Raton, FL, 226 pages.

2. Kelly, Donna, & Rupert, Edwin. 2009. Professional Emotions and Persuasion: Tapping Non-rational Drivers in Health-Care Market Research. *Journal of Medical Marketing: Device, Diagnostic and Pharmaceutical Marketing* 39: 3–9.
3. Cunningham, Ceara Tess, Quan, Hude, Hemmelgarn, Brenda, et al. 2015. Exploring Physician Specialist Response Rates to Web-Based Surveys. *BMC Medical Research Methodology.* doi:10.1186/s12874-015-0016-z.
4. Pit, Sabrina Winona, Vo, Tham, & Pyakurel, Sagun. 2014. The Effectiveness of Recruitment Strategies on General Practitioner's Survey Response Rates—A Systematic Review. *BMC Medical Research Methodology.* doi:10.1186/1471-2288-14-76.
5. Wadhwa, Himani Singhi. 2016. Global Markets and Technologies for Advanced Drug Delivery Systems. www.bccresearch.com/market-research/pharmaceuticals/advanced-drug-delivery-systems-tech-markets-report-phm006k.html (Accessed January 2, 2016).
6. Fuld, Leonard. 2006. *The Secret Language of Competitive Intelligence*, Crown Business, Boca Raton, FL.

Chapter 3
Situation Analysis

The preparation of a marketing strategy is not done in a vacuum. It is the result of a careful analysis of an organization's current situation, from both internal and external perspectives. As such, situation analysis is the first step in building a marketing strategy. We will be breaking down situation analysis carefully in the next few pages, emphasizing the difference between internal and external analyses.

Internal analysis enables the company to identify its own strengths and weaknesses. Internal elements are those which the company has created and that it can control. These include existing processes, capabilities, and resources as well as the organization's vision and business objective. Of specific interest for companies in life sciences is the selected business model, as this has a direct impact on how a company will address its customer base and its competitors as well as look for potential partners.

External analysis is related to elements outside the organization, which directly impact a company's operations. As external elements are outside the organization, the company (as an individual organization) has very little control over them. These factors include your customers, your competitive environment, the market scope (size, forecasted growth) as well as the environmental analysis (which includes the political, environment, social, and legal environment).

The last few pages of our chapter will be dedicated to how to classify these elements using a simple model, SWOT (strengths–weaknesses–opportunities–threats), followed by a brief demonstration on how to use this data to build a more actionable model, TOWS (threats–opportunities–weaknesses–strengths) matrix.

> **On the importance of assessing your own situation**
>
> When we look at a new company presentation, there are two key elements we come to expect. First, the company has to demonstrate the impact of the proposed technology on patient health and on disease management; this demonstration has to be clinical of course, but the impact has to be demonstrated from the commercial point of view as well. For example, a technology could work from a clinical perspective, but not make any commercial sense, and be impossible to commercialize. Second, the company has to be able to demonstrate its differentiation to existing technologies: is this a me-too product, or something that reshapes the market? In both cases, are you able to articulate the value proposition of your product in the market?
>
> *Marc Rivière, General Partner and CMO at TVM Capital*

3.1 Internal Analysis

When performing internal analysis, you will be looking at three distinct aspects of your organization. First, you will review the company's mission and vision. Following this, your analysis will focus on your current assets (such as your physical and human resources, your capabilities, and your core competencies). Finally, in life sciences, the business model will need to be well defined, as it will have a considerable impact on the marketing strategy.

3.1.1 Assessing Your Corporate Vision and Mission Objectives

A company is more than a product or a service. It is the result of an entrepreneur's dream (or a group of entrepreneurs, sharing a common vision). This vision can range from the very broad ("Make the world better") to the more specific ("A computer in every house").

If your company already exists, you should already have a mission and a vision, even if it is not formalized. A clear understanding of your organizational mission and vision is important to keep in mind when you develop your marketing strategy, as the two elements provide direction and a clear sense of purpose for those inside the organization. If you do not have a corporate vision and mission statement, it can be very tempting for an organization to become overly opportunistic and pursue opportunities outside your core business.

Building a company's mission and vision

Here are a few tips to prepare and write a good mission statement and a good vision statement for those who do not already have one.

A company's **mission** is the expression of the intent of the organization. It defines the goals and objectives of the organization in a short and succinct manner. Your mission statement should answer three key questions:

1. Who is your intended target client?
2. What product or service will you provide to that client?
3. What makes your product or service unique?

> A good way to verify your mission statement is to ask yourself, "Can any of my competitors use the same mission statement?" If the answer is yes, then you have not found yet a mission statement that uniquely fits your organization.
>
> As for your company's **vision**, it describes the desired future outcome of the company and helps people outside the organization understand what your company is all about. Ideally, it should describe a desired result and should evoke emotion. It should also be short: try to limit yourself to one or two sentences, and like the mission statement, try to avoid generic terms and statements. If it can be applied to other companies, then you still have not found your unique company vision.

3.1.2 Assessing Your Current Capabilities

Making an inventory of your current capabilities will be useful to understand what your organization can currently accomplish, as well to identify gaps. In a start-up situation, most of your current assets will be found around the expertise and network of the founders, but could also include access to partners, funding, intellectual property (IP), or past experiences that can be leveraged in this current endeavor. To assist in this assessment, you can use the checklist of elements provided in Table 3.1 to help you evaluate your current organizational capabilities.

During this assessment phase, it is important to emphasize the current assets which generate a sustained advantage. It is by focusing on those assets that you will be able to distinguish yourself going forward.

In situations where the data is available, a company can also engage in performance analysis, measuring multiple data points such as customer satisfaction, quality of its product (perceived and real), brand image, and costs. Benchmarking these numbers against industry standards will enable the

Table 3.1 Checklist to Assess a Company's Current Capabilities

Capabilities	Questions to Ask
IP	1. Do I have IP protection around my product(s)? 2. Can my product(s) be protected by IP? 3. Is there an advantage of having IP protection? 4. Do I have an IP strategy?
Physical resources	1. Do I have access to the physical resources that I need (lab space, lab equipment, ingredients, etc.)? 2. Do I have access to any specific resources that are not available to competitors, giving me a competitive edge?
Human resources	1. Do I have access to specific personnel to cover the whole chain of the business opportunity, or are there gaps? 2. Are my personnel invested in my organization, or is there a risk that they could leave for a competitor? 3. Do I have access to unique personnel/key opinion leaders (KOLs) not available to competitors?
Funding	1. Is my current funding sufficient to reach my business goals? 2. Do I have access to funding that will enable me to reach my business goals? 3. Can I leverage current funding to generate more funding?
Product	1. What is the development stage of my product? 2. Are product derivatives possible? 3. Is this a product that has unique attributes?
Cost	1. Do I have a source of cost advantage, such as economies of scale, experience curve, or product design innovation? 2. Is cost advantage (or cost parity) essential to my product marketing strategy?

company to get an appreciable image of itself vis-à-vis other comparable companies. We will be going in detail on various performance indicators (such as return on assets and market share) in Chapter 6.

3.1.3 Assessing Your Company's Business Model

In life sciences, a company's business model is incredibly relevant to its marketing strategy. There are many different business models a company can adopt, and it is important to decide early which type of model it will use. As such, the company must ask itself questions such as the following:

- What is the goal of the company?
- Is the market reaching the goal?
- Is the goal to develop a functional product, to demonstrate the product's market viability, or to generate revenues?

This goal must be as clear and single-minded as possible.

In therapeutics, there are a number of models a company can adopt. Basically, the company has to decide how deep in the commercialization chain it wants to invest in, if it wants to cover multiple aspects of the regulatory chain, or if it plans to develop a specialty in one specific space.

For discussion purposes, we can define the major steps to reaching the market across two different tracks. One concerns the scientific and product track, whereas the other follows the commercialization track (Figure 3.1 and Tables 3.2 and 3.3).

Let us take a moment to examine some of the more popular business models in life sciences such as the fully integrated model, the virtual model, the platform model, and the no research/development only model.

Figure 3.1 **Major steps toward commercialization in therapeutics.**

Table 3.2 Major Steps toward Commercialization in Therapeutics—Scientific Track

Step	Definition
Discovery and development	**Discovery** is the identification of therapeutic candidates which can be used for medical treatment. **Development** identifies how the best candidates are absorbed by, distributed through, metabolized in, and excreted from the body; what is the best possible dosage; and the best delivery system.
Preclinical research	Before testing a product on humans, it is tested in vitro (on specific cells), in vivo (on animal studies), and in silico (through computer simulations).
Phase I	Phase I studies are held on 20–100 healthy volunteers to determine optimal dosage and establish safety.
Phase II	Phase II studies are completed on up to several hundred people with the disease/condition to measure efficacy and monitor side effects.
Phase III	Phase III studies are made on anywhere from 300 to 3,000 volunteers who have the disease/condition to measure efficacy and monitor adverse effects over longer periods.
Manufacturing	Manufacturing encompasses all steps relative to manufacturing a product, including production, packaging, and distribution of the final product.
Phase IV	Once a drug is approved for sales, companies are responsible for continued monitoring of their drug. This long-term surveillance is made to ensure that the drug is safe with a larger group of patients.

The classical model in therapeutics is the **fully integrated model**, where the company attempts to integrate all business functions under a single corporate umbrella, from discovery

Table 3.3 Major Steps toward Commercialization in Therapeutics—Marketing Track

Step	Definition
Market Analysis	Pre-commercialization research to understand the market, its trends, and the competing companies; done to ensure that there is a market opportunity for the product or service
Commercial planning	Preparation relative to the launch of the product on the market. Includes commercial strategies and more operational decisions (such as selecting the distribution network and setting the pricing policy)
Prelaunch activities	Marketing campaigns, strategies, and methods you use to get the word out about your new product or service, targeting potential buyers to get them excited about your upcoming product
Product launch	Official launch of your product on the market, the first time your clients can officially purchase the product
Commercialization	Ongoing activities (from advertising to customer support) to ensure and increase the sales of your product/services, converting interested parties from followers to customers

and development to manufacturing, sales, and marketing. This model is also referred to as a "bench to market company." Some companies generate revenues by licensing out a few compounds, and then selecting a few other products to commercialize them themselves. This was once a more popular model for start-ups, but with time, investors and management teams have found that there are a number of issues with this model. First, it is quite costly to own and operate equipment from across the spectrum, to fund all the clinical trials, and requires many types of distinct expertise (from researchers, to production, to managers and sales staff). Furthermore, companies that operate in this model carry heavy fixed costs. While it is less likely today for therapeutic start-up companies to aim for an integrated model, some of the original big biotech companies (such as Gilead) are from this mold. If companies in the nutraceutical and medical device space are somewhat likely to utilize this model due to a shorter clinical approval process, the fixed costs and dilution of expertise concerns remain.

A more popular model right now in therapeutic companies is the **virtual model**. These are companies which, quite early, decide which segment of the value chain they have expertise in, and then develop the company to fill the specific function. Usually, these companies focus on the discovery and development functions. As such, a virtual life science company could decide to focus on the discovery and development, out-licensing products as these get regulatory approval, and starting research again on new products. Having a virtual model limits costs and risks, as you usually have less fixed cost, reducing your burn rate.

The **research model** is a variant of the virtual model. Rather than bringing the drugs through clinical phase, these companies focus exclusively on the research and preclinical phase, licensing out compounds to bigger companies that can then focus on the clinical phase. This is done to allow the company to contain costs even further, but the potential royalties of licenses are much lower. This is the type of company that often comes out of academia, funded by government grants, and often has only a single product in its pipeline.

Platform companies are another variation of the research model company. In this case, a company has access to a proprietary platform that it uses for the identification of therapeutic target. During the first phase of its existence, the company focuses on partnering its platform with companies to generate revenues, shifting to building its own pipeline once it has sufficient capital and traction.

Another therapeutic model is the **no research/development only** model. In this model, a company focuses exclusively on the clinical development of products. It in-licenses products that have cleared the preclinical phase and focuses on bringing them to the end of the phase III, where it will out-license them to big pharmaceutical companies. It usually generates revenues through royalties and licenses.

Other more specialized models include the **recovery** model (focusing on products that failed advanced clinical studies due to weak efficacy and repurposing them for new indications), the **combination** model (combining two drugs for new indication, which are then sold to big pharmaceutical companies), and the **drug delivery company** (which focuses on developing a novel delivery mechanism for existing drugs).

In diagnostics, companies use various models as well:

1. They can decide to develop their diagnostic product and license their technology for manufacturing and commercialization, akin to the research model.
2. They can decide to license the test to an established diagnostic company, so it can be paired and included in an existing device.
3. They can decide to go to market with their own diagnostic device and commercialize it (akin to the fully integrated model).

As for medical devices, they often use business models similar to therapeutic companies, but with a shorter regulatory path and a shorter path to market. As such, it is much more common for medical device companies to favor fully integrated models, or to reach the commercialization stage (including the manufacturing step) and look for a partner to handle distribution and sales, as commercialization of their product is a must for partnering.

Indeed, in one of my interviews, Lidija Marusic, investment manager at Innovacorp, who specializes in medical device life sciences investments, mentioned that she had noticed very different partnering strategies between big pharmaceutical/ biotechnology and big medical technology companies. "With therapeutics, partners typically buy-in early and work to bring the product to market, whereas with medical devices, you have to take the product to market: big medtech companies need to see some initial market traction, unless the product is really innovative."

Overall, your decision will come down to choosing how long you will work on the product internally and how much of the marketing you will ultimately be responsible for. But even if you decide to focus on research and development exclusively, it will be crucial for you to do some of the market planning to demonstrate the viability of your product.

3.1.4 Some Final Notes on Internal Situation Analysis

When doing your internal situation analysis, it is possible that some of the content that you generate is of limited value. Some common mistakes include listing too many elements (or being too detailed), or some of the elements look less like facts and more like opinions.

If you have added too many elements in your analysis, it will be confusing and difficult to get a strong picture of your key attributes at a glance. To resolve this, try going through your list again and remove some of the elements you feel are more peripheral. Getting a set of fresh eyes to look at them independently might be a good way as well to remove some elements which you listed as critical, but do not hold to scrutiny.

Another recurring issue is that some factors that you will identify might look more like opinions and less like facts. This can occur when the internal situation analysis was performed by a single individual, without the feedback from a second person. Once again, getting another colleague to look over your internal situation analysis and having them evaluate it and give feedback would be a good way to sift through fact and opinions.

3.2 External Analysis

External analysis is the analysis of factors that are outside the organization, directly impacting its functioning. While the company (as an individual organization) usually has very little control over these factors, it is very important for it to understand them. Furthermore, a company can attempt to influence some of these factors by collaborating with other companies within its ecosystem through a chamber of commerce or an industry association, for example.

For this section, we will be going over customer analysis in life sciences, the competitive environment, and the market scope (size and forecasted growth) as well as the environmental analysis, which includes both the microenvironment (Porter's Five Forces) and the macro-environment (using the Social, Legal, Economic, Political, and Technological [SLEPT] model).

3.2.1 Customer Analysis

Your customer is the individual or organization that will be **purchasing** your product. The goal of the organization's marketing strategy is to **attract** the customer, **gain** his attention, and **convert** this attention into a purchasing decision.

As we will detail in the following pages, there is often a disconnect in life sciences between the customer (who pays for the product), the consumer (who uses the product), and the purchaser (who decides which product will be purchased). Sometimes, the same individual fills the same role, whereas in other situations, up to three different individuals interact in the same transaction.

Understanding your customer is key to being able to convert his interest into a purchasing decision. As such, we will start by going over the classical model of the customer decision-making process, followed by a short overview of the specific customer ecosystems for the life sciences industry, a short section on identifying consumer needs, and finish this section with a few words on consumer behavior.

3.2.1.1 Understanding the Customer's Decision-Making Process

The best way to address the customer's needs is to start by understanding his decision-making process. Understanding the driving forces of your current and potential customers' purchasing process will help you better identify marketing opportunities and optimize your marketing strategy.

There are five steps in the classical customer decision-making process.* First, he recognizes that he has a problem and that he has an unmet need. This is followed by a search for information, which is then followed by an evaluation of different solutions. The fourth step is the actual purchase of the selected solution, and the last step is post purchase actions (such as product reviews or repeated purchases). To properly illustrate this process, we will use a very simple example in life sciences, where the patient has a headache, and looks to solve his problem (Figure 3.2).

First, as in any customer decision-making process, he recognizes and identifies that there is an issue. He might identify a specific unmet need to be filled, or he might recognize that the current product he is using is problematic and doing its task very poorly. In our example, the problem is that the patient has a headache (or could be that he is using a

---

* Note that newer customer models eschew the funnel customer purchasing process and believe that engaging the client is more of a cyclical process rather than a funnel, adding more importance to the feedback stage as an opportunity for repeated purchases. Others subdivide the steps for a more complex and thorough model. But for our current needs, the classical model is sufficient to elicit thought for the entrepreneur.

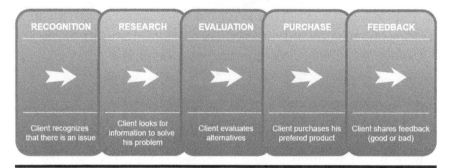

Figure 3.2 The classical customer purchasing process.

headache medication which does its job poorly due to a slow action mechanism or secondary effects, for example).

Of note, it is also possible to elicit and stimulate an unmet need in consumers through advertisement and other media tools. For example, if you have developed a headache medicine that also simultaneously helps the user sleep, and then develop an advertisement campaign promoting that specific combination of characteristics, it might awaken a reaction in the client, who might never have realized that he needed to have both solutions in the same product. Targeting users that are at this stage is ideal as you get to the forefront of the user's decision process. An easy methodology for identifying the most important client needs is the Kano model, which we will be reviewing in Section 3.2.1.3.

Following the recognition of a problem or issue, the customer will start looking for information on potential solutions to help him decide which product to purchase. He can engage on a wide range of activities. Some of his research can be accomplished by contacting individuals who are more knowledgeable than him (doctors, nurses), or people who have previously used the product or alternatives (such as friends and family members). He can also look for information on websites, magazines, and online reviews by using popular search engines. In our simple headache medication example, a visit to his local pharmacist and a couple conversations with his friends and family would probably be the extent of the research he would perform. This is the stage where a company typically engages and advertises to potential clients, when they are actively looking for information. The company will typically try to reach them through advertisements, promotions, and digital media.

Customers then evaluate the different products they have identified. They might be using a more formal decision process (e.g., using a "pros" versus "cons" process to compare different products) to manage risk. The main questions the customer asks himself is, "Is this the product I need?" and "Will this product solve my problem?" If the answer is no, then he goes back to the second phase, looking for more information and alternative solutions. Hence, the process is iterative, and it is possible for the client to cycle through the second and third steps many times.

Once a decision is made, the customer purchases the selected product. If you are responsible for managing the purchasing process (through your own website, or through your own assigned staff), make sure your purchasing process is simple and efficient; there is no better way to frustrate and disenfranchise a client than by making him go through a clunky purchasing process, and then lose him as he is unable to complete the transaction.

The final step in the customer purchasing process is the review process. At this stage, the customer might review the product, leaving feedback for other users. This is especially true if you used a third-party platform for your sales, such as Amazon. If he was particularly impressed by the product, he will share his feedback to other potential users. Note that the number of people leaving reviews will be pretty low: on average anywhere from 1% to 3% of consumers will leave feedback on a product.

It is also possible to recruit participants who have had especially positive experiences as brand ambassadors, who will then go on to interact with other potential users. These brand ambassadors have many names in life sciences such as technology champion or super user. They are sometimes remunerated or compensated for their work, depending on what is expected of them. Hence, some companies implement brand ambassador programs to recruit, coordinate, and reward these individuals.

Finally, in many industries, marketers who put a lot of emphasis on branding, believing that customers will often favor products they are initially aware of, have a more favorable impression during the research phase. This is less of a factor in life sciences, as end users often favor elements such as efficacy over image. In a recent paper, marketers found that skin care consumers were purchasing as many products they identified during the research phase (38%) as those that they had identified in the initial consideration phase (37%) [1]. Indeed, as the costs in life sciences are constantly rising, customers are increasingly sensitive to the value and prices of products in life sciences, opting for generic products as they are unable to identify clear advantages to buying premium branded products. Some ways to fight against private labels include investing in brand equity (through continuous product improvements), avoiding the launch of fighting brands (positioned between private labels and noted brands), managing the price spread, and exploiting sales promotion tactics [2].

3.2.1.2 The Client Ecosystem

A seemingly simple question with a very complex answer:
Who is your client?

Life sciences companies interact with a complex client ecosystem as a cornucopia of actors and agents can be involved in the purchasing process. This complex ecosystem leads to situations in which the consumer of your product is not necessarily your customer. Moreover, in some situations, the person paying for your product might not even be the decision maker (he might not even be the client). Hence, for some products, there is a disconnection between the consumer (the person who will use your product), the purchaser (the person who will decide which product to purchase), and the customer (the organization who will be paying for the product).

Unfortunately, too many times, this disconnect will have a deep impact on innovation, and new technologies will be unable to integrate the market.

To illustrate the complexity, let us start by a very simple non-life sciences example. Let us say you are selling a pair of gloves. You advertise said gloves on television. The client, after viewing your commercial, decides to visit your website to purchase them or visits a local store to purchase them directly. In this situation, the consumer, the purchaser, and the customer are all one and the same.

Now, let us make the process a bit more complex. Say you are selling surgical gloves used in operating rooms. The end users are surgeons and nurses in the operating room. So, do you advertise to them? It depends. On one side, it is highly unlikely that they are going to be the individuals paying for this type of product. So, while they might not purchase gloves directly, they might be able to influence the internal purchasing processes. But already, there is a disconnection between the consumer and the customer, while the decision-making responsibility could be shared.

Now let us make our example a bit more complex. Let's say you have designed a diagnostic machine for lung cancer that completes a diagnostic in 5 minutes but costs $100 for each time it is used. Current hospital costs are $10 for each diagnostic, but it takes 12 hours to complete. For the patient, the convenience is immeasurable, but since he is not the person purchasing the device, he is not privy to both options and he is not necessarily the customer (his insurer might be paying), his needs and opinion are trivial compared to the saving being generated to the hospital through the traditional method.

As such, there are four main relationships in life sciences purchasing process:

a. *The relationship between the consumer and the purchaser*: A patient (consumer) needs a medication for his condition. His doctor (purchaser) tells him to use brand ABC rather than using another brand. The patient decides to purchase it.
b. *The relationship between the consumer and the customer*: A patient (consumer) needs a tooth implant. The insurance company (customer) only reimburses bridges. The patient chooses to use this procedure instead of a full implant.
c. *The relationship between the consumer and the customer*: A doctor (consumer) recommends changing from one medical supply product to another. The Group Purchasing Office (GPO—customer) acquiesces, recognizing the doctor's expertise on the topic, and purchases the new product instead.
d. *The relationship between the consumer, the purchaser, and the customer*: A doctor (consumer) prefers a brand of surgical gloves. The hospital (purchaser), which pays for the product, while sympathetic to the doctor's preference, is constrained by his GPO (customer) to purchase another brand.

In the text that follows, we will be going over the main stakeholders implicated in the purchasing process, their role, and their specific interest: the patient, the hospital, the caretakers, the private insurer, the GPO, the hospital, the primary caregiver, and the government agency. It is very important to understand how your product is purchased and the relative influence of each actor on the purchasing decision.

In many cases, the **patient** is the end user/consumer of a product. As such, he is often perceived as a decision-maker, and the target of initial marketing strategy from the perspective of the product being sold to him. For example, products that perform diagnostics quicker are often believed to be invaluable for the hospital setting. But, unfortunately, patients' preferences often have very little impact in health-care settings on what product will be purchased. As such, if your product is not being sold directly to the patient itself, positive attributes that benefit the patient often have less value. You should be highlighting key attributes of cost saving and time saving, rather than focusing on the client's comfort. If your product is directly targeting the end user (for usage and payment), attributes such as pain relief and health benefits all have tremendous value.

Primary caregivers (such as doctors and nurses) can also play a key role in adopting new products in a health-care setting. As such, these stakeholders can act as champions of technology, interacting with hospitals to share insight into what products are worthy of considerations. Nonetheless, doctors are not always decision makers in hospital settings, so while a product that makes their job easier might draw some interest and support, if it is costly compared to the current gold standard, it could be difficult to gain traction. As such, doctors and nurses can play a role in technology and product adoption, but they are not always decision makers since they are neither a direct beneficiary of the technology nor the party paying for it.

In some health-care systems, **private and public insurers** play an important role in health care as they reimburse part or the totality of a treatment, a product, or a device they have specifically selected. As such, products which they refuse to cover often have tremendous difficulty integrating health-care markets. The role of insurers will vary a lot from one situation to another, but from a general perspective, it is important to understand that insurance companies have their own evaluation processes to select which services and products they will reimburse, and will usually favor those that reduce costs. The products that are covered vary from one company to the next, and biological processes (which are especially expensive) are not always covered. Finally, remember that insurance companies will only cover the cost of a product or service they have initially approved, so if an alternative product is used, they might only cover the cost of the equivalent product, or even nothing at all.

A **GPO** is an organization combining a number of like-minded organizations, leveraging a combined group purchasing power to lower the price of a product or service due to higher expected volumes. The objective is to generate savings for its members, aggregating large volumes. One of the main issues when dealing with GPOs is that their involvement can add some level of rigidity to purchasing procedures. In some GPOs, memberships are constraining agreements: if the GPO negotiates an agreement to purchase a specific product, all members will be constrained to purchase the same product. GPOs will often play a role in selecting products for its members, and as such, it will be important to get a read of the environment and see if your type of product is subject to purchasing restrictions. As the number of GPOs increases, pricing is quickly becoming one of the key attributes when selecting products.

Hospitals are organizations dedicated to giving care to patients in need. They can be for-profit or nonprofit. Depending on their relationship with GPOs and insurance companies, they can have various levels of latitude when selecting products. Either way, they will often have extensive purchasing processes and a department dedicated to the procurement of products, selecting products and services based on request for quotations.

The **primary caregiver** is not a stakeholder that will be present in every health-care situation, but when he is present, he becomes an important influencer in purchasing decisions. A caregiver is usually defined as someone from the patient's network who helps the patients with daily activities. The objective is to help preserve the persons' autonomy and ensure that they can recover. Caregivers are usually involved in situation of long-term care, disability, or old age. They will usually be sensible to services and products that can assist them in their tasks, which might not immediately benefit the patient. Products that help track patient movements around a house, or reduce burden when doing recurrent tasks (measuring heart rates automatically) are simple examples of products that could target caregivers.

Finally, there is the role of the **government**, which varies greatly from one location to another. It will often also have the role of regulator and can also impact the prices through reimbursement lists. In some situations, it will be the purchaser, and it will have the same role as a GPO, putting pressure on prices through consolidated purchases. In some locations, it will favor the purchase of local products as a way to encourage the local business cluster.

On the importance of understanding the purchasing process

Having multiple stakeholders implicated in the purchasing process can make it challenging for start-ups looking to accelerate product adoption. In these situations, the best approach is to identify the stakeholder who has the most impact and concentrate efforts on convincing him first. In some situations, it might be the hospital administration. In others, it might be doctors, so it's important to understand the purchasing process correctly. Once this stakeholder is onboard, proceeding top-down to other stakeholders is more efficient than dealing with multiple vectors simultaneously.

Katherine Parra Moreno, Vice President of Business Development, Epic Capital Management

3.2.1.3 What Does My Customer Want?—Using the Kano Model to Understand Your Customer

Understanding what your customer needs the most can be a daunting task. Yet, during your product development, it is essential to prioritize your efforts on essential features. An interesting model to use for this is the Kano model of attribute classification.

The Kano model facilitates the classification of customer perceptions, and it enables decision makers to identify which features will generate higher returns (since these create higher satisfaction) and which features are not worth investing in (since they are not creating any satisfaction, or even worse, generating dissatisfaction).

There are five categories for customer purchasing attributes: threshold attributes, performance attributes, excitement attributes, neutral attributes, and reverse attributes.

A. *Threshold attributes* are those the customer expects in a product to fulfill his basic needs. They must be present, and since the customer anticipates them, they usually do not create a differentiation opportunity. Increasing the focus or investments on these attributes does not generate more sales or greater client attraction, but removing or diminishing these attributes will result in customer dissatisfaction. A simple example of a threshold attribute is the absorbent pad on an adhesive bandage. The client expects there will be an absorbent pad on a bandage; the lack of a good absorbent pad on an adhesive bandage will turn clients away, yet it would be difficult to focus on this pad as a unique selling proposition.

B. *Performance attributes* are those which, the more they are present, the better it is for the customer, improving customer satisfaction. They are sought out by the client. The price for which the customer is willing to pay for a product is tied to performance attributes. Nonetheless, products that focus on performance attributes are interchangeable

with other products focusing on performance attributes, so it is difficult to build a strategy that focuses exclusively on this attribute as a product strategy. An example of a performance attribute in an adhesive bandage would be the glue which makes the adhesive stay on the finger longer.
C. *Excitement attributes* are unanticipated by customers, but generate high levels of satisfaction. However, their absence does not generate dissatisfaction. These are the attributes that enable you to differentiate your product from products with similar performance attributes. Providing excitement attributes that address unmet needs generates a competitive advantage over competitors. Following our bandage example, bandages with antiseptics integrated into the absorbent pad is an example of a product with excitement attributes for customers.
D. *Neutral attributes* are attributes for which the customer does not have a distinct opinion. Their presence neither improves nor impedes the purchasing decision. Frequently, these are features that excite the entrepreneur, but seldom build interest in customers. While it does not mean the attribute does not add value to the product, it just does not add any value for the customer. For example, the length of the packaging envelope for a bandage is seldom evaluated by customers, but reducing its length might make it easier to manufacture and package.
E. *Reverse attributes* are rather rare and occur when a client perceives an attribute negatively. In that case, he will be willing to pay an amount of money to have them removed. As such, the presence of a reverse attribute creates a negative experience, while its absence creates a positive experience. For example, some clients might be willing to pay more for plain adhesive bandages, rather than buying cheaper adhesives bandages with kids' cartoons on them.

The Kano model is a simple yet powerful model, useful to identify opportunities. The more the attribute creates customer satisfaction, the more effort should be invested in developing and refining it. Nonetheless, there are two issues with Kano modeling. One is the impact of time on attributes, and the other is related to divergent consumer perceptions.

The main issue with Kano modeling is the impact of time on customer perception. An attribute that was initially classified as an excitement attribute can quickly become a threshold attribute as more companies integrate the target feature. For example, think back a few years ago about when cameras and portable music players were integrated into mobile phone devices. If it was once an excitement attribute (*Wow! I don't have to carry three devices anymore; everything fits onto a single device!*), could you now imagine purchasing a mobile device that could not play music, or could not take pictures? As such, the Kano model needs constant reappraising to ensure that attributes are correctly classified, and that the market has not shifted away from some of your previous assumptions.

The second issue is that you can find inconsistencies among customer perceptions. As you start collecting data, you might realize that while some attributes are threshold attributes for some individuals (a camera is a must for some potential clients), they become negative attributes for others (*I have to pay extra for a camera? But I never use it! Show me a cheaper model without a camera, instead!*). Careful segmentation of your data alleviates this issue: As you identify precisely your customer segments, you will eventually distinguish patterns between them. The clearer your end market, the less likely you will have inconsistencies in consumer perceptions.

Overall, the Kano model is a useful model to identify your key product attributes and to plan carefully your commercial activities according to customer needs and expectations.

3.2.1.4 Identifying Customer Behavior

Customer behaviors can be classified into four main categories, and identifying which one applies the most to your

Situation Analysis ■ 53

customer will play a huge role in the development of your marketing strategy:

A. *The economic model*: In this model, the customer always attempts to purchase the least expensive product and always works toward maximizing its own benefits. As such, strategies that focus on establishing a competitive pricing and maximizing the consumer's purchasing power will influence how you interact with your client. GPOs and hospitals that are operating in cost-based assessment systems are typically included in this type of consumer behavior.
B. *The learning model*: In this model, the client will buy products based on immediate needs and the learning process that occurs after the purchase. For example, if a client is ill, the client will buy the best medication, and would not necessarily compare quantities and online reviews. The immediate need (pain resolution) determines the product purchased.

   Furthermore, if the client is rewarded following the purchase (e.g., the client is satisfied by the level and speed of pain relief), there is a strong tendency to repeat purchases of the same product. The more satisfied the client grows, the more loyal the client becomes. Inversely, if the client is unsatisfied, he will stop purchasing the product. Usually, patients and caregivers who are directly involved in the purchasing decision as well as the payment of the product usually fall in this category.
C. *The psychoanalytical model*: It sees the purchasing decision as a combination of conscious and unconscious elements. As such, subjective elements (such as branding elements—packaging, color, taste) play a huge role in purchasing patterns. Individuals are most likely to follow this consumer behavior, rather than organizations.
D. *The sociological model*: Fundamentally, purchasing behaviors are based on the peer group of the purchaser. Hence,

the group has a lot of influence on how customers make purchasing decisions.

3.2.1.5 Building Customer Profiles

Customer profiles are fictional personas that you build to represent typical customers purchasing your products. Building these profiles helps you better understand your customers, what makes your product unique for them, and is useful for preparing your marketing strategy, developing your advertising collaterals, choosing your distribution channel, and so on. Each time you are preparing your tactics, customer profiles become a useful sounding board against which to check your assumptions.

For example, if you just prepared an ad, you can take a second look at it through your customer's eye, asking questions such as the following:

- Would my customer like this ad?
- Will it make him look for more information?
- Will it be a call to action?

Building personas will require an in-depth understanding of your clients. This will mean performing secondary research, supplemented by one-to-one interviews and site observation if that is possible. You want to get as acquainted as possible with your shortlisted best potential clients.

Once your information is collected, you can start building the persona. Some key information you typically compile and develop includes the following:

- Typical title/role in the organization
- Goals and motivation
- Challenges and pain points
- Where do they learn about new products and services?
- What is the best way to engage with them?

- What is their purchasing process?
  - Do they have purchasing authority?
  - What is the purchasing customer behavior?

For example, let us go through a theoretical persona. In our scenario, the company has developed an innovative technology that is used in lab research and is looking to better understand its potential customer.

Using the profile found in Figure 3.3., we can see some of the current pain points are around sensitivity and required supervision. Also, engagement is best performed by on-site visits and demos, rather than doing web-based contacts or traditional advertising. Finally, purchases are done based on cost, and while Terry—The Typical Lab Tech is not the decision maker, she is an important influencer (Figure 3.3).

3.2.1.6 Some Final Notes on Customer Behavior

One aspect of customer behavior that should not be discounted is customer inertia. Some customers will have purchasing habits that do not fit the models and processes we have listed earlier and will demonstrate considerable "loyalty" to their favorite products, even if the product shows considerably worse characteristics. This is somewhat difficult to evaluate, but as a rule of thumb, you should remember that most consumers are conservative, avoiding new products to remain with known products. They will usually change if they do not have the choice to do so (current product price has risen dramatically, it is no longer available, etc.), or if they are encouraged by a third party (referral from a friend or family member, for example).

Customers usually display weak customer loyalty for small purchases, for products with less added value, or for those done without much forethought. For these products, it is easier to change the purchasing habit, but as the customer's

> **Terry—The Typical Lab Tech**
>
> **About**
> - Lab technician in a university research center
> - Has been in the same position for the past 4 years
> - Performs experiments in the lab and handles day-to-day tasks
> - Uses technologies that are currently available in the lab, not too keen to learn new technologies
>
> **Goal and motivation**
> - Wants to successfully complete her experiments
> - Needs to ensure results are accurate and reproducible
> - Prefers simple technologies, even if they take more time
>
> **Challenges and pain points**
> - Technologies currently used have sensitivity issues, making her redo experiments often
> - Some technologies require intensive supervision, reducing her ability to multi task
>
> **Where does she learn about new products and services?**
> - Typically relies on vendors for information on new technology
> - Does attend two to three conferences per year, but does not focus on new technology
> - Reads generic science magazines (*Nature, PubMed*) and those relevant to her specific field of research, as well as online media
>
> **What is the best way to engage with her?**
> - Likes it when vendors come to her lab with the "new toys," so she can try them out
> - Frequency has to be approximately once a month, so she can dedicate half-hour sessions
> - She discards generic emails and pays more attention to phone calls and personal visits
> - Advertising has little to no impact, but social media has been steadily more important
>
> **What is the purchasing process in her organization?**
> - New purchases must be approved by the lab director
> - While she does not have the purchasing authority, she usually makes recommendation for purchasing
> - Purchases follow the economic model — Cost is very important when choosing products, so she has to make a solid case around potential savings
> - Products that save time have little chance of being approved

Figure 3.3 Example of a customer persona: Terry—The typical lab tech.

investment in the product increases (in terms of time and money), so does the brand loyalty.

Finally, if you lose clients after a single purchase, or if the customer turnover is considerably elevated, consider preparing a voice-of-the-customer* initiative to identify the motivating problems and causes of dissatisfaction.

---

* A voice-of-the-customer initiative is a project where a company solicits feedback and information for its client in the form of surveys and interviews, trying to identify opportunities for improvements.

3.2.2 Competitor Analysis

Founders often believe that they have developed a unique product or service which will revolutionize the market, and as such, they do not need to perform a formal competitive analysis. From their perspective, their product is so unique, uses such a unique technology, or delivers value in such a unique way that no other product can compete. Research seems to confirm this observation: studies tend to demonstrate that when founders are asked to identify competitors, they usually think of a relatively small number of companies and use supply-based attributes (who they are and what they do), rather than look at demand-based attributes (what customers want and need).

Nonetheless, one of the basic tenants of competitive analysis is that all products and services have competition, and it is unrealistic to believe that there is no competition to a product. There is always an alternative. It might be indirect, it might be costlier, but somewhere, there is a product or service that is competing in a very similar space than you.

To challenge this potential myopia, competitors should not be defined in terms of the technology. Instead, they should be defined in terms of the consumer need they are fulfilling, and which aspect of that need is currently unmet. For example, a diagnostic test for a condition might already exist, but there might be an unmet need on the ability to share results or the ability to integrate a secondary test at the same time.

> **On the importance of identifying the correct opportunity**
>
> When looking at a company, it's not just about the technology. It's about the team, their vision, their strategic plan, the opportunity they identified and how passionate

> they are. Sometimes the technology fails, but the company still succeeds. A great team has the ability to pivot, develop or license in a new technology and refocus the company towards a successful path.
>
> *Paulina Hill, Principal at Polaris Partners*

Once you have defined your market from the perspective of "what need is my product answering?" and "how is my market currently answering to this need?" rather than focusing on the technology, the competition field will be clearer.

Also, it is important to refine the notion of a direct competitor versus an indirect competitor. A direct competitor is a company that is answering a need in a way that is virtually the same way as yours, whereas an indirect competitor is one that is answering a need, but in a completely different way. For example, if you open a clinic catering to clients with skin cancer, direct competition would be other clinics, while indirect competition could include apps that detect skin cancer through a phone and platforms that offer virtual doctor consultations online.

In this section, we will be going over how you can identify and evaluate your competition. We will describe four simple steps of competitive analysis: preparation, identification, evaluation, and modeling.

3.2.2.1 Preparation

During the preparation stage, you take a moment to decide three things: the goal of the competitive analysis, the definition of the product or service you will analyze, and the perspective that you will use to complete your competitive analysis.

  a. *Goal of the competitive analysis*: It is very important to articulate the goal and objective of the competitive scan.

Is it for internal or external purposes? If the analysis is mostly for external purposes (third parties), it will be crucial to identify and reference the sources you consult. It will also be crucial to determine the scope of the analysis: How wide and how deep do you wish to go during research? It is important to set expectation early (i.e., identify the top five competitors) rather than start research, and stop once you "feel" you have done enough research.

b. *Definition of the product or service you wish to analyze*: Take a moment to define the competitive space for your specific product and the need (either filled or unmet) that you are targeting. This cannot be underestimated, and can be illustrated in a simple example I worked on a few years ago.

At that time, my client was developing a health-care application in the self-care space. One of the issues was that the definition of the app and what it could do shifted from one conversation to the next. While we had carefully defined what information we wanted to collect, the definition of what the product did, and what it could do shifted constantly, making it hard to identify competitors: competition identified in one iteration was not necessarily applicable to the next iteration.

This is an issue which is quite common in marketing strategy and market research. What comes first, the product or the market research? In some cases, this is extremely simple, since the product is defined by its technology and development phase (what it does is determined early in its design phase). But sometimes, a product will have flexible end points (this applies a lot in health-care IT and medical applications), or will have multiple target market opportunities. As such, product definition becomes purposefully vague and is expected to be formalized during market research. To resolve this issue, some constants are necessary before starting market research and situation analysis. A good compromise is to do iterative phases, using the constant elements as the basics for your next research phase. We will be coming back to this later when we discuss the product-market fit in Chapter 4.

c. *Decide on the perspective you wish to take*: Your competitive research can be built from an inside-out perspective or an outside-in perspective.

In the **inside-out perspective**, the company is looking from inside the company toward the market, looking at other competitors through its own lens, through their own perspective and metrics, to identify what is important to them. What are their perspective, their strategy, and their approach toward clients? In other words, you define your competitors the way you believe that they are approaching the market, making parallels with your own organization.

In the **outside-in perspective**, you look at your competition, not from your own perspective, but rather from the perspective of your potential client. Hence, you attempt to get into your client's mind-set when he is going through his purchasing process, and try to replicate it. Doing so enables you to identify what might be appealing to the consumer and what is important in your customer acquisition and retention strategy.

For example, a few years ago, a client wanted me to identify all his competitors for his product which was a medical device with high regulatory hurdles. After some research, it became clear that purchasing a product in another country was terribly prohibitive, and the number of actual companies that were active in his market was much less than he initially believed. As such, options for his potential consumers were limited to a small number of specific companies, not necessarily all companies producing a product that answered the same need around the world.

3.2.2.2 Identify Key Competitors

The second step is to identify the key competitors. You might believe that you already know this; most likely, you already know some of your competitors that you often see, cross, and compete with, but it is unlikely that you already have a complete threat assessment.

As such, it is important to identify as many companies as possible that could be filling the same need as yours. During this research phase, you could identify the same company from multiple sources. Note this information. This could be indicative of the companies' relative importance in the current competitive ecosystem. Your competitors include other start-ups, larger established organizations with competing products, all the way to universities incubating early technologies. Also remember to be quite open-minded on how you define a competitor. You can always weed out superfluous targets later in the process. Compile this list of competitors, sorting out duplicates, and you will be ready for the evaluation phase.

During this phase, you will also have the opportunity to identify the "gold standard." This product will be the one that is most commonly used, and against which new products are benchmarked in terms of the impact on both human health and cost. As such, correctly identifying the gold standard will allow you to make a realistic assessment of the additional value your product provides and to evaluate how likely the purchasers are to evaluate your product.

3.2.2.3 Evaluate Your Competitors

There are several approaches that you can take to evaluate your competitors. We will go over two of them: building competitive profiles and creating a competitive array. The advantage is that both methods are complementary, and it is quite useful to engage in both exercises.

3.2.2.3.1 Competitive Profiles

The first methodology is to build competitive profiles of the most relevant competitors. You might already have some idea of the most relevant competition by the information density around the data gathered. Companies mentioned more frequently are more likely to be actively pursuing the same market, while a company mentioned once or twice might be dormant or inactive, or might have exited the market. Investigate these further, focusing on signs that confirm inactivity: for example, no employees on LinkedIn, information only found on a government registration site, no news presence, and lack of management or employees associated with the company (beyond the initial founder). If you want to be thorough, you might want to set up a Google Alert if any activity occurs around this company, or contact the company directly to assess its level of activity (see Section 2.4.2 for more information on setting up a Google Alert).

You might also want to inquire with colleagues about these outliers. A few years ago, a client had contracted me to build competitive profiles for companies developing diagnostic solutions for Alzheimer's disease. I had identified a small company in a provincial directory, but could not find any supplementary information. As I presented information to my client, he was quite amused, as I had found information on a former company related to his current company, which had been closed over 5 years ago, but which was still listed in online regional directories.

Building competitive profiles has a number of advantages. First, by doing this, you might identify gaps in the current market and opportunities you can exploit. Also, by structuring the data this way, it will be easier to update the information through regularly scheduled research efforts.

Each competitive profile sheet will be adapted to your situation. Here is a generic one you might be able to adapt to your needs (Table 3.4).

Table 3.4 Sample Competitive Profile Checklist

Company Name	Competitor #1	Competitor #2
Headquarters		
Website		
Estimated revenues		
Company description		
Current management		
Management names and background		
Quality of top and middle management		
Knowledge of business		
Entrepreneurial culture		
Loyalty/turnover		
Innovation		
Internal R&D capabilities		
R&D partnerships and alliances		
Product pipeline		
Patents		
Marketing		
Unique selling proposition		
Current commercial strategy		
Commercial partnerships and alliances		
Distribution		
Sales forces		
Customer services/product support		

(*Continued*)

Table 3.4 (*Continued*)   Sample Competitive Profile Checklist

Company Name	Competitor #1	Competitor #2
Manufacturing		
Internal capabilities/capacity		
Outsourcing strategy/model		
Finance		
Current estimated revenues		
Current estimated cash on hand		
Major financial partnerships and alliances		

3.2.2.3.2 Competitive Array

The second technique is to build a competitive array. The advantage of this tool is that it enables you to identify which companies are the biggest threats, and which are less of a threat. There are three simple steps to building the competitor array, which you can start once you have built your customer profile.

1. Step 1: Codify the key success factors in your industry, as perceived by your end user.
2. Step 2: Rank those factors by weighing them, once again through the perspective of your end user.
3. Step 3: Rate your competitors. If possible, have two or more people go through this rating exercise using the same grid, then collate the information.

You might be asking yourself a few questions at this point:

- *Why am I using the customer's perspective to evaluate competitors, rather than using my own?* Quite simply, during the purchasing process, the customer is the one

evaluating one competitor against another. He is the one doing the purchasing, not you, and as such, your perspective is less useful.
- *How do I get my customer's perspective?* There are a number of ways to do this. Sometimes through direct conversations with them (formal or informal) at trade shows, events, or conferences, for example. It is also possible though customer feedback forms, voice-of-customer initiatives, or focus groups. Another way (although less reliable) is through review of online discussion boards, product reviews, and forums.

You can then calculate a weighted score for each company/product and use the information when preparing your marketing strategy. As an example, imagine a medical device company that has completed its customer profiles and has successfully identified its customer's needs. Through competitive profiling, it has identified that company A has produced a product that focuses on the quality and customer experience, whereas company B focuses on pricing and ease of use. So, which one is the biggest threat? (Table 3.5).

Table 3.5 Demo of a Weighted Score Evaluation of Competitive Space

Key Industry Success Factors		Company #1		Company #2	
Component	Weight	Raw Score	Weighted	Raw Score	Weighted
Price	0.3	4	1.2	8	2.4
Quality	0.2	8	1.6	4	0.8
Delivery to consumer	0.1	8	0.8	4	0.4
Ease of use	0.4	4	1.6	8	3.2
Total	1	24	5.2	24	6.8

Earlier, in its research, the medical technology company had identified price and ease of use as the most important attributes for potential customers. As such, using ponderation/weighted scores, the company identifies company B, which is most closely aligned to target customer needs, making it the most important competitor.

3.2.2.4 Getting Information on Competition

There are multiple resources you can use to both identify and evaluate your competition. Here are a few resources to check for both the identification of key data and indications for more in-depth research.

3.2.2.4.1 Private Company Websites

Competitor websites will often have valuable information on the company's technology, management, and marketing strategy as well as their own perceived benefits. Some companies will post their pitch deck from previous roadshows, whereas others will post annual reports, analyst reports, media articles, patents, corporate brochure, and more. Competitor websites are also invaluable to learn how a competitor positions itself or to learn what the company perceives as its unique selling proposition.[*]

Remember to be critical of the information you find on competitor websites, as companies have all the incentives in the world to exaggerate claims. The information you find on these websites can be useful to triangulate further research, or identify the opportunities for further research. For example, a statement of a new collaboration with a university might enable you to find a university portal for research centers, chairs, and small start-ups in the same space as you.

---

[*] We will be coming back to the notion of the unique selling proposition in Section 4.4.1.

3.2.2.4.2 Advertising Materials

A company's advertising materials include both online materials (online ads, sponsored content) and physical materials (brochure, magazine ads). Typically, a company will emphasize its own strengths and downplay its weaknesses in its own marketing materials. As such, getting your hand on its promotional documentation (which are public documents) can be invaluable to understanding its market perceived position, as well as comparative data it might have compiled.

3.2.2.4.3 Online Media

Online media (both generalist and specialized) often run stories on different companies in the life sciences space, and it can be very interesting to review them to have access to competitors' position, as well as analysts and thought leaders. These sources will often highlight information such as partnerships, fundraising, acquisitions, and management changes. You might even gain access to information such as market data, financials, and other quantitative information.

3.2.2.4.4 Conferences

Conferences are great sources of information. Companies often have speakers presenting on panels and forums, and during these presentations, you will have access to some privileged information as these companies share information to the audience.

3.2.2.4.5 Patent Databases

Using a patent database is a great way to identify competitors that are developing similar technologies. As such, you might be able to identify a company's initiative using pertinent keywords. Some patent databases you might wish to investigate include the United States Patent and Trademark Office (patft.uspto.gov) or the European Patent Office (www.epo.org).

> **On the importance of understanding the competition**
>
> When we evaluate new opportunities, one of the things we take a look at is how it is going to replace the current standard of care. As such, in initial conversations, entrepreneurs have to show a good comprehension of how the product is going to be used, what the market is currently using, who the main competitors are, and what the unmet need that the innovation is targeting is.
>
> *Lidija Marusic, Investment Manager, Innovacorp*

3.2.2.5 Final Notes on Competition

Remember that you should look at your competitor not necessarily in terms of how the technologies compete, but rather how they answer a client's needs. Your main competitors are those who are competing along the same client's need, not necessarily those who have the same technology. Also, building that customer profile will be key in understanding how to properly evaluate competitors and will provide valuable information in building your marketing strategy.

3.2.3 Market Analysis

Market analysis is done to describe the market that the firm is actively targeting. For the purposes of this book, we distill market analysis into three components: market size, market forecasting, and market trends.

3.2.3.1 Market Size Estimation—The TAM-SAM-SOM Model

As a market researcher, one of the top questions my clients have is "What's the market size for my product?" When doing these projects, my favorite model to use is the TAM (total addressable market)-SAM (serviceable available market)-SOM (Serviceable Obtainable Market) model.

The main strength of this model is that it combines both the top-down market sizing approach (assessing the market size from market research) and the bottom-up approach (estimating the market size from the perspective of the organization's available resources).

There are other advantages for this model. First, it is relatively easy to use and explain. Second, it combines the top-down and bottom-up systems of market research that are essential to telling the entrepreneur's story. Finally, a well-built TAM-SAM-SOM model demonstrates a solid understanding of the market and its limits.

The biggest limit of this model is that it relies on the strength of the entrepreneur's assumptions. As such, building a solid TAM-SAM-SOM model is reliant on a realistic assessment of the market and on the capabilities of your organization. Furthermore, the justification behind your commercialization plan will be crucial in justifying your SOM.

3.2.3.1.1 The Three Parts of the TAM-SAM-SOM Model

The TAM-SAM-SOM model is divided into three distinct parts that are imbricated one into another. The largest part is the TAM for a specific product or service. It contains the SAM for the market that is within the reach of your company. This, in turn, contains the SOM, which is the realistic market share that you can obtain, or the one that you plan to achieve in the short term (Figure 3.4).

3.2.3.1.2 TAM-SAM-SOM Example

To illustrate properly, let us use the very simple example of an imaginary start-up called MedLung, which is developing a diagnostic device for small-cell lung cancer.

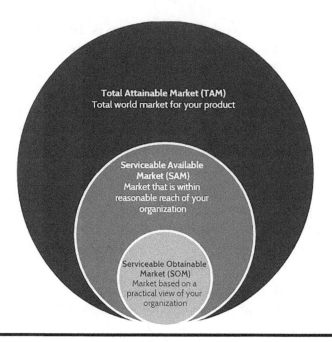

Figure 3.4 TAM-SAM-SOM model framework.

The TAM is the market from the broadest perspective. For our small start-up, the TAM is the total lung cancer diagnostics market space. Grand View Research estimated that the global lung cancer diagnostics market revenue would reach US$ 3.6 billion by 2024.* We can use this as a starting point for our future world TAM.

But MedLung cannot conceivably claim this total market, so the next step is to calculate the SAM, which is the market that is within MedLung's reach. In this example, we work using two assumptions:

a. The diagnostics device MedLung has developed specifically detects small-cell lung cancer, which accounts for ~15% of the lung cancer diagnostics market share.
b. In the short term, MedLung plans to target the North American market exclusively, which accounts for 30% of the revenue share.

---

* Source of the estimated market size.

Table 3.6 Sample Serviceable Available Market Development

Estimated TAM	US$ 3.6 billion
Estimated market share of small-cell lung cancer segment	15%
Estimated market share of North American market	30%
Estimated SAM	US$ 162 million

Hence, MedLung's SAM is estimated to be US$ 162 million (Table 3.6).

Finally, MedLung needs to determine the SOM, which is the realistic market share that it can likely acquire. This estimate has to be backed by a bottom-up forecast, which details how resources will be utilized to achieve these revenue goals. Without going into the details of a commercialization strategy, let us imagine that MedLung has a commercialization plan that generates US$ 10 million in sales. Data in hand, MedLung can demonstrate quite efficiently its market size (both current and potential) using a bubble graph as shown in Figure 3.5.

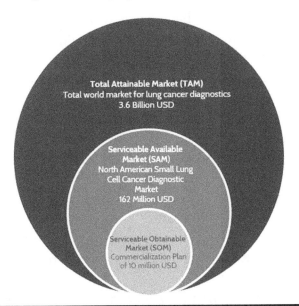

Figure 3.5 TAM-SAM-SOM model for a hypothetical small-cell lung cancer diagnostics device.

You might have noticed that this type of modeling depends on the quality of data and the hypotheses you make. As such, your market research will be the cornerstone of your end results. A good understanding of your market and its competitors is essential to building a justifiable SOM. Remember that recipients will be as interested in the numbers as in how you developed the model: Was the company realistic when building its model? As a rule of thumb, forecasting a SOM which is greater than 10% of the SAM is going to require a great deal of justification.

Also, as with other models, TAM-SAM-SOM models are iterative and can be adjusted as new data become available. Finally, it is always a good idea to be a bit more conservative with your numbers. It is easier to explain how you beat your forecasts, than to justify why you fell short of the number promised.

> **On the importance of realistic assessments**
>
> Ultimately, the value that you generate is driven by your technology's actual impact on the healthcare system as a whole. Unfortunately, some companies have "non-actionable discoveries", or technologies that will be very challenging to commercialize. Faced with this challenge, a few companies make unrealistic claims, overestimating the marketing potential, or trying to link their product to what needs to be marketed (rather than emphasizing their own technology). This is a constraint of the commercial innovation, it needs to be applicable immediately. Nonetheless, these companies are not making themselves any favors when overestimating their market and impact as we identify these exaggerations quite effectively.
>
> *Marc Rivière, General Partner and CMO at TVM Capital*

3.2.3.2 Market Forecasting

There are three main categories of forecasting methodologies: judgmental methods, cause–effect methods, and market research methods.

Judgmental methods generate data based on the opinions of consulted experts and KOLs. The expertise, opinions, and judgments are then turned into a formal prediction. This methodology, while preferred by managers, is often critiqued as being subjective and inaccurate. Judgmental methods include the following: (1) Delphi, (2) expert judgment, (3) scenario writing, and (4) decision trees/assumption-based modeling. These techniques are relatively easy to use on a small data sample and are useful to forecast for new and innovative products, but they can generate biased predictions (due to the effects of group dynamics, for example), they can be expensive and time-consuming if done on large groups, and some of the people building the estimates can have a vested interest in overestimating or underestimating their forecast.

Companies that have already existing sales can use cause–effect methods, which are based on using historical sales data to project forward. If you have access to reliable data, this method is inexpensive and objective, but you must be careful of making incorrect assumptions based on cyclical or seasonal date, human error, or product life cycle. Also, this approach leaves very little room for judgmental or non-recurrent/exceptional events. For example, it would be difficult to use this model in a situation in which you are subject to an exceptional event (such as a one-time large purchase order). Finally, this method is not very useful when investigating new and emerging products.

Being of a market research background, I am biased toward the use of market research methods, especially in a start-up situation. To use these methods, data is gathered by exposing potential consumers to the new product and then measuring their purchase intention. It is also possible to do a test market assessment in which the product is launched and measured in a controlled and limited environment.

Forecasting based on user expectations is done by surveying potential clients. As such, projected product interest can be calculated accurately and can be used for new product forecasting. The main issue is that the accuracy of your forecast will depend on the predictions of the individuals you survey, and if you have contacted overly positive individuals, it could lead to overly optimistic sales estimates. Also, this method can be labor-intensive and time-consuming. Finally, as surveyed individuals are not committing to the purchase, it is usually prudent to attenuate sales estimates to account for "polite" participants (those who answer in the positive more in an effort to answer what they think you want to hear, rather than relative to actual purchase intent).

As for market test, the objective is to replicate the purchasing sales process in a limited environment and to measure success, extrapolating results to a larger market. For example, a company could start making the product available to clients in a city, limiting marketing activities to this market, and then extrapolate sales for bigger markets. The advantage is that you are testing the real market and the reactions of consumers to your product. When forecasting from expectations, your forecast will be biased as people are estimating their purchases, but when doing a test market, people are "putting their money where their mouth is." Also, it allows the company to assess the effectiveness of its marketing programs. The main downsides are that it makes competitors aware of your product prematurely, is quite expensive to set up, and can take a lot of time to measure results.

Using the data that you gathered during your market test and survey of customer expectation, you can build hypothetical sales funnel, which you can use to forecast your revenues in a comprehensive and systematic methodology.

The sales funnel is based on your hypothetical sales process. In our case, let us imagine a company having a simple sales process:

1. **Qualified leads**: A prospect which you have identified as a good fit following the completion of your customer profiles. You should use a checklist to qualify leads and discard those that do not fit your ideal customer profile. Churning through unqualified leads can be a waste of time, as these leads can drain time through your funnel all the way to qualified prospect.
2. **Prospect**: A potential client who has demonstrated interest in hearing more about your product.
3. **Qualified prospect**: A potential client that has demonstrated interest in purchasing your product, can authorize the transaction, and has committed to the purchase.
4. **Committed**: A potential client that has accepted your offer and is proceeding with the purchase.

It is important to understand the notion of conversion rate: Not every single contact that displays interest will convert into a sale. Hence, the conversion rate is the metric you will determine to describe the probability of a contact going from one step of the sales process to the next. Let us imagine a scenario in which, during our customer survey/market test, you found that

1. Fifty percent of potential clients surveyed expressed interest in purchasing your product.
2. Twenty percent of potential clients that you met to pitch your product expressed interest in purchasing it.
3. Fifty percent of potential clients expressed interest in purchasing or preordering your product (i.e., committed funds).

Table 3.7 Customer Funnel Table Sample

Stage	Number of Targets	Conversion Rate (%)
Lead	100	50
Prospect	50	20
Qualified	10	50
Committed	5	—

You could then build the product table (Table 3.7).

In short, this would mean that for every 100 leads you identify, 5 are converted into a purchase. Using bottom-to-top forecasting, you might determine that your sales staff would be able to reasonably manage a number of targets per cycle, and then calculate your potential sales.

Finally, remember that every company's sales funnel is unique; they all have different stages and shapes. Some might have extra steps where there is an external assessment/due diligence step, where there is a set meeting process with associates, a test with end users, or more. While it is fine to start with a generic sales funnel to forecast sales, it will be much more accurate once you have customized it for your unique customer requirements and processes.

3.2.3.3 Final Notes on Market Analysis

Beyond the marketing models, do not forget to take an intuitive look at your market. If there are a large number of potential users, there is high growth in the number of potential users, and it is easy to acquire new clients, you have a good market on your hand. Inversely, if your market is small, not willing to spend, and shrinking, it will be that much more challenging.

3.2.4 Environmental Analysis

The last step of the situation analysis is to correctly identify your company's environment. This will be essential to understand where there are opportunities within your business environment and to better understand the micro and macro-trends that are going to directly impact your organization.

During your environmental scan, you will analyze your competitive environment from two different perspectives: the micro and macro perspectives. From the micro perspective, you will focus on your organization's immediate environment such as competitors and collaborators to effectively position yourself. To do this, I suggest starting with Porter's Five Forces framework. From the macro perspective, your organization will take a wider look at its environment, determining how macro-level dimensions (policy, economy, etc.) affect your company. For innovative companies, I suggest using the SLEPT model, which is a variation of the more traditional PEST (political, environmental, social, and technology) model, but which includes a legal component.

3.2.4.1 Microenvironment—Porter's Five Forces

Porter's Five Forces is a simple yet efficient framework that you can use to model your immediate environment. Due to its popularity, it will be recognized easily by most stakeholders. Porter's Five Forces are the bargaining power of suppliers, the bargaining power of buyers, the barriers to entry, the threats of substitutes, and the overall industry rivalry (Figure 3.6).

- **The bargaining power of suppliers**: If you are in a situation in which there are few product suppliers, or you are using a very rare component, there is a risk of being faced with several issues such as being charged exaggerated prices or having supply problems. This will severely impact the growth of your company and your ability to manage this growth. Conversely, the more suppliers are available, the less they will have a hold over your company, and the more freedom of action you will have to grow your company.

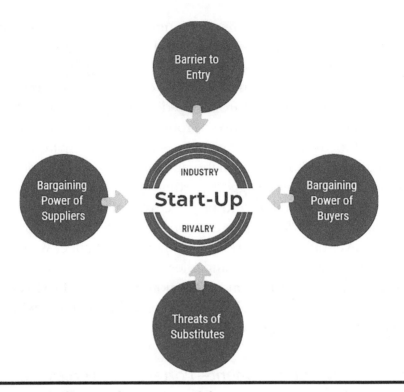

Figure 3.6 Illustrated Porter's model.

- **The bargaining power of buyers**: As with suppliers, the more amalgamated your purchasers are, the more power they will have over your company. If your product is sold to a limited number of GPOs, and you have few alternative clients, then your existing clients will have a lot of influence over your organization. Alternatively, if you are selling one of the few products available on the market, and consumers have almost no alternatives, then you will have more influence over your consumers. Other elements to consider include the relative strength of your brand and the cost to switch from your product to alternatives.
- **The barriers to entry**: Barrier to entry can make entry into your market more difficult for new competitors. For example, if your product needs a specific certification to be sold, then the number of new entrants will be

considerably lower. Other barriers to entry can include economies of scale, access to distribution, IP, capital requirements, and switching costs. An important barrier to entry to be remembered in health care is switching costs. These are the one-time costs that an organization must make to replace the existing infrastructure. For example, in intravenous (IV) solutions and kits used in hospitals, procedures for attaching solutions to patients differ among competitive products, and the hardware for hanging the IV bottles are not always compatible. In this case, switching encounters resistance from nurses responsible for administering the treatment and requires new investments in hardware [3].

- **The threats of substitutes**: Substitute products are products that address the same needs as your products. For example, if you are a contract manufacturer of solid-state drugs, substitute products could include ocular drug delivery systems and gum-based delivery systems.
- **Industry rivalry**: The level of competitiveness versus collaboration plays a huge role in assessing the environment. In some industries, companies will dedicate resources to increase collaboration rather than competing, focusing on creating integrated and interdependent value chains. You should also consider industry capacity, growth, scope of fixed costs, and industry concentration. The faster the market is growing relative to your capacity, the less inter firm competition there will be as companies struggle to keep up with overall demand.

Some types of competition, notably price competition, is prominently unstable and is quite likely to leave the industry worse off from a profitability perspective. Price cuts can be matched quite rapidly by competitors, while more long-term strategies, like those tied to advertising, can create more differentiation and enhance the brand.

Finally, you should remember that Porter's Five Forces model was developed in the late 1970s. As such, it has drawn criticism today as being overly static, unable to account for disruptive change. It is also criticized for placing too much emphasis in the competitive aspect of the environment, not accounting for strategic alliance and inter-collaboration between companies. Finally, as this is a micro-level framework, it does not consider some of the more macro factors (such as government regulation or societal changes). These will be investigated in the SLEPT model.

Nonetheless, like many other frameworks, Porter's Five Forces enables you to quickly identify the information you need to collect and lets you present it in a systematic fashion that is understood across multiple stakeholders. When it is used by itself, it can be useful to give an organization an overview of its situation and begin its situation analysis until more complex models are used.

3.2.4.2 Macro-Environment—The SLEPT Model

To analyze the macro-environment, we suggest using a simple framework, the SLEPT model. While some stakeholders may be more familiar with the more classical model of PEST, it is essential to add the legal component within the framework of a small start-up in the life sciences space, due to the overall importance of intellectual property. Hence, the model breaks down a firm's external environment into five simple components: social, legal, economic, political, and technological. In the text that follows, we will be going over each factor, and then presenting simple methodologies to complete your own SLEPT model (Figure 3.7).

The five components of a SLEPT model are defined in Sections 3.2.4.2.1 through 3.2.4.2.5.

3.2.4.2.1 Social

Social components include the behaviors and lifestyles of your clients, as well as their demographic profiles. They are especially

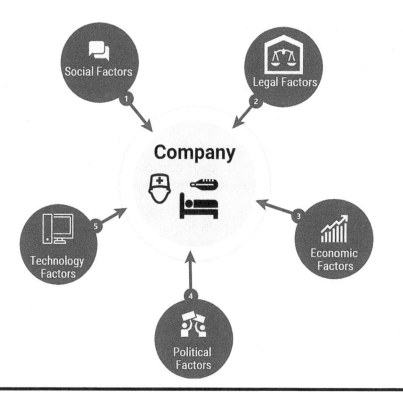

Figure 3.7   Illustrated SLEPT model.

important if your company has a strong business-to-business component (i.e., if you sell mostly to other organizations). If you are constantly in contact with your end users/clients, and they are directly implicated in the purchasing decision, the social environment will be very important. Aging populations are often a key component to a company's SLEPT model in life sciences.

3.2.4.2.2 Legal

The legal framework is a key component in life sciences as it plays an important role in incentivizing innovation. Protecting your innovation from competitors will mean figuring out if you will use an IP strategy, trade secrets, or a mix of both. Beyond IP, the complexity of the legal framework (employment and consumers laws) and the regulatory framework will play a large role in determining how a company operates and can commercialize its products. For example, in a large number of countries, companies are not allowed to advertise certain types of health products directly to consumers, whereas in others, it is possible if the ads are preapproved or even without any form of pre-approval.

3.2.4.2.3 Economic

This factor is related to the performance of the economy where you wish to establish your company. Is the economy growing, or is it in a downturn? Other factors can include exchange rates (especially if you are purchasing components abroad) or interest rates (which impact the cost of raising/borrowing capital).

3.2.4.2.4 Political

The political aspect includes how the different levels of governments (national, state/provincial, municipal) are interacting with companies in your space. Have they put into place strategies to fund, retain, and support this sector, or do they view this sector of activity as ancillary? Governments can impact a sector of activity through a variety of levers, including taxes and tax breaks, subsidies, legal framework, and policies. On a secondary level, macro decisions that the government makes (such as those on health, education, and infrastructure) can all impact your company.

3.2.4.2.5 Technology

This will often include elements such as the rate of research and development investment and technology incentives. Understanding how technology advancement impacts your organization is a key component of your own marketing strategy.

3.2.4.3 Building a SLEPT Model

The best way to build a SLEPT model is to go through a planned brainstorming session, and then research the gaps that you identify. Identify key leaders in and around your organization to participate in your first brainstorming session to ensure you have identified all relevant trends and issues for each of the topics. To assist you in the process, we have included five simple questions for each component to help guide initial discussions.

3.2.4.3.1 Social

1. What is the target's population growth rate and age profile? Is this likely to change? How?
2. What are some of the top lifestyle trends, attitudes, and opinions affecting the target market?
3. Has there been a shift in consumer buying patterns which could affect the organization (e.g., credit, virtual currencies, online purchases, and co-payments)?
4. Are there generational shifts in these trends between age groups that are likely to affect the company?
5. What is the media's attitude toward the company's sector of activity? Is there a difference between local, national, and international perspectives?

3.2.4.3.2 Legal

1. What is the current legislation environment in the target market? Any future legislation which could impact it?
2. How well developed are IP rights in the target market? Is IP protection enforced?
3. Are there any specific rules or regulations around advertising and publicity in the target market for the product/service?
4. Is there heavy supervision from home regulatory processes and bodies for the sector of activity? Does the company have a good grasp on the regulations that directly affect the product/service?
5. How widespread are corruption and organized crime in the home/target market? How are these situations likely to change, and how is this likely to affect our organization?

3.2.4.3.3 Economic

1. How stable is the target economy? Is it growing, stagnating, or declining?
2. How stable are relevant neighboring economies? Are they growing, stagnating, or declining?
3. Are key exchange rates stable, or do they tend to vary significantly?
4. Are key interest rates stable, or do they tend to vary significantly?
5. What is the unemployment rate? Will it be easy to build a skilled workforce? Is there a shortage of qualified workers in the target location?

3.2.4.3.4 Political

1. When is the country's next local, state/provincial, or national election? How could this change government or regional policy?
2. Who are the most likely to obtain power? What are their views on our sector of activity, and on other policies that affect your organization?
3. Could any pending legislation affect the organization, either positively or negatively? What is the probable timeline of proposed legislative changes?
4. Are there any political policies that support our sector of activity?
5. Are there any international legislation/trading policy/funding grant changes on the horizon which could impact the organization?

3.2.4.3.5 Technology

1. Are there any new technologies that could affect the organization or the industry?
2. Are there any new technologies that the organization could be using?
3. Do any competitors have access to new technologies that could redefine their products?

4. In which areas do governments and educational institutions focus their research? Is there anything the organization can do to take advantage of this?
5. Are there existing technological hubs that the organization could work with or learn from?

3.3 Classifying Outputs: From SWOT to TOWS

Once you have gathered your information, it is time to classify it into a framework. To do this, I suggest completing a SWOT framework, and then converting the information found in the model into a TOWS matrix, a useful model to identify the actions to take in your marketing strategy.

3.3.1 The SWOT Model

The SWOT model is a simple yet clear framework which can be used to present your information. In this model, the strengths and weaknesses will refer to the organization itself (internal elements), whereas opportunities and threats will focus on the environment outside the organization (external elements). Hence, you will be using the information you obtained in the first part of the analysis to fill the strengths and weaknesses part, and data on competitors and the environment to fill the opportunities and threats part.

The main advantage of the SWOT model is that it is easily recognizable: most individuals have been confronted at one time or another with the model, and a quick glance will usually indicate to participants the nature of your model, letting them capture the information you are conveying. The popularity of the model lets the user condense and simplify complex data. This is its disadvantage, as a tendency to oversimplify complex situations can leave information off the table. Another issue is that every factor is weighted equally in a SWOT model, so it is impossible, at a first glance, to determine which factors are the crucial ones and which ones are ancillary. Finally, SWOT models are subjective, so expect the building of your SWOT model to be iterative, changing as more information becomes available and more feedback is obtained.

3.3.1.1 The Four Elements of a SWOT Model

The four elements of a SWOT model are split into two broad categories: internal elements (strengths and weaknesses) and external elements (opportunities and threats). In general, internal elements are those that you have some measure of control over, whereas external elements are those which you have no control over. As a reminder, internal elements were examined in Section 3.1, whereas external elements were reviewed in Section 3.2 (Figure 3.8).

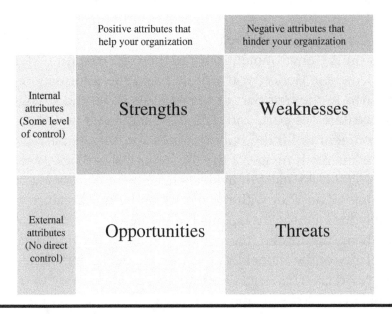

Figure 3.8 SWOT model framework.

Strengths are the positive elements that give your organization a competitive advantage or unique value over competitors. A specific technology, an exclusive license, and a strong patent portfolio are all unique attributes that can serve to distinguish you from existing competitors. Team members with unique skills and experiences are also strengths that need to be showcased.

Weaknesses are the attributes that are inhibiting growth in your organization. They are the attributes that place you at a disadvantage compared to your competitors. For example, using a legacy production system to build your product, lack of sales expertise, or not having a patent to protect your innovation are all weaknesses that can inhibit your growth.

Weaknesses can be difficult to define and admit. Many organizations are wary of facing their own internal weakness, or believe that drawing attention to these weaknesses will be a potential strike against them during investor pitches. The important thing to remember is that defining weaknesses is the first step, but the important information to share is how the company will address its weaknesses. As such, recognizing and addressing a weakness is a much better alternative to hiding it under the rug, hoping nobody will notice it.

Opportunities occur when there are changes in your environment that you are uniquely positioned to take advantage of. New regulations, exclusive access to new suppliers, or the difficulties of an important competitor all generate opportunities for your organization. Opportunities can emerge in technologies, social and lifestyle trends, population changes, government policy, markets, and more. As such, continuous monitoring of your environment is invaluable to identifying opportunities.

Threats are shifts outside your organization that can impact you. While they are outside your organization (meaning you have little control over them), recognizing and assessing them can help you prepare your organization, and identify solutions to address them. For example, a shift in regulatory procedures could incentivize you to change some elements of your production model, and look for appropriate subsidies and partners to help you make the necessary adjustments.

You might not be able to change the external environment, but you can react once the change occurs (or even better, you can plan ahead to make adjustment before the changes occur). Some examples of threats include new competitors entering your market, current client shifting their strategy, new technologies shifting consumer purchases and use of your products, changes in social trends, government regulation, or changes in supply of crucial materials.

3.3.1.2 Developing Strategy Applications— From SWOT to TOWS

One of the main criticisms of SWOT is that it is a static framework and does not showcase how each factor you identified relates to others. It does not turn into action, being a rather still frame of what is going on. For example, you might have identified a specific strength which can compensate for a specific threat, but a SWOT table would not illustrate this.

That is why using a TOWS matrix is so interesting, as it lets you use the data collected in your SWOT to turn it into a strategy, an active element for your marketing plan. To do so, you build a matrix inside the matrix. The key becomes understanding how data that you obtained from one field interacts with data obtained from another field. You cross the data as shown in Figure 3.9.

This creates four quadrants where you develop specific strategies:

  a. *Matching strength and opportunities (SO)*: This is also called *matching* and implies the pairing of your positive attributes (strengths and opportunities) to identify the prospects for your organization. For example, if one

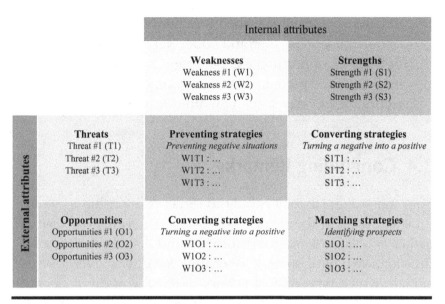

Figure 3.9   From SWOT to TOWS.

of your strengths is a unique production process, and an opportunity emerges in lower regulatory hurdles for a specific product, you might identify that product as a new market opportunity as you are able to scale up rapidly.

b. *Matching weakness and threats (WT)*: This is also called *preventing*, as you try to minimize weaknesses and avoid threats. These strategies will be defensive in nature, with the objective of protecting yourself from potential threats. When preventing, you will not be identifying opportunities for new successes. For example, if your product requires expensive components, and you expect a higher competitive environment due to fluctuating currency, a defensive measure could be used to increase the purchase of less expensive foreign components as a way of matching prices.

c. *Matching strengths and threats (ST)/weakness and opportunities (WO)*: Also called *converting*, this activity involves identifying a weakness or threat and transforming it into a strength or opportunity. For example, if a competing start-up is developing an innovative technology which threatens your market, then it might be an interesting opportunity to partner and codevelop said technology. Other opportunities could be developing a similar innovation or partnering with a competing innovative technology to develop a new unique selling proposition.

3.4 Concluding Remarks

Your marketing strategy is built in direct relationship to your understanding of the market. As such, the information you gather is invaluable to going forward and making the right decisions. We have spent a lot of time for this chapter to explain which type of data is important, so you can collect the data you need to formulate your strategy. Understanding your environment, your competition, as well as the trends shaping your market will be essential in moving forward.

References

1. Court, Dsavid, et al. 2009. The Consumer Decision Journey. *McKinsey Quarterly*, www.mckinsey.com/business-functions/marketing-and-sales/our-insights/the-consumer-decision-journey (Accessed December 18, 2017).
2. Quelch, John, & Harding, David. 1996. Brands versus Private Labels: Fighting to Win. *Harvard Business Review on Brand Management* 74(1): 99–109.
3. Porter, Michael. 1998. *Competitive Strategy, Techniques for Analyzing Industries and Companies*, The Free Press, New York, 10 pages.

Chapter 4

Developing a Marketing Strategy

Developing a marketing strategy is an extensive venture, so we will be dividing it into multiple sections. Section 4.1 will deal on topics relevant to your marketing strategy vision. This will be followed by a short discussion (Section 4.2) on traditional company marketing models, and which one is the most relevant for you. As such, you will be connecting the information you gathered in Chapter 3 with the type of company you wish to plan for. Sections 4.3 and 4.4 will delve into the more practical applications of marketing strategy and describe fundamentals such as the pricing strategy distribution strategy, and promotional strategy. Finally, due to its importance, we conclude with a distinct section on digital marketing aspect (Section 4.5), while adding more practical elements throughout the chapter.

4.1 Selecting Your Marketing Strategy Vision

The bulk of this chapter will be dedicated to going over the different marketing tools and applications that you can use. It will be a lot more practical. But before we get to that point, let us take a moment to discuss the overall strategic vision that will influence these strategic decisions.

The three main strategic visions are strategic commitment, strategic opportunism, and strategic opportunity. These are not mutually exclusive, as one might best apply to one aspect of your business and, while not apply to another. As such, it is possible to have a mix of all three simultaneously in the same company, as long as each one is applied to a distinct company function/product. For example, it is possible for a company to pursue a vision of strategic commitment when it comes to its distribution and its product strategy, while having a more opportunistic vision relative to promotion-related decisions.

4.1.1 Strategic Commitment

The *strategic commitment* vision involves establishing a clearly defined marketing strategy, and then choosing actions which will ensure you attain these goals. It is a long-term vision of your organization, and forgoes short-term opportunities that might emerge midstream in favor of attaining the long-term goal. The "sticking to your guns" strategy ensures that your focus and resources stay fixated on the long-term goals ahead. It also provides your whole organization a common vision, ensuring everybody is on the same page.

This does not imply a static and inflexible application of your strategy. Your company has to be able to adapt to external and internal events as they occur. But this does mean that changes will need to be structured as improvement to your current strategy, rather than completely changing the end goals. Hence, the objective is the continuous improvement of the strategy rather than the creation of a new one each single time an opportunity emerges.

The main risk linked to this philosophy is the appearance of *strategic stubbornness*. This occurs when a company persists in continuing forward with a strategy when something in its environment has changed, making the current strategy a wasteful exercise. The classical example of this is a company continuing to sell and innovate in the typewriter market when personal computers are obviously displacing them. Being unable to recognize and adapt to external and internal changes can often offset the benefits of unity of vision that are obtained by a strategic commitment.

Making faulty assumptions of the future could also undermine your marketing strategy significantly. For example, I worked on a project for a company developing an innovative biomass conversion technology. When developing his marketing strategy, one of the major assumptions that the client had made was made relative to partnerships. He believed that the components used in conversion, which where waste by-products, would stay free in the foreseeable future. After all, companies were paying to have them removed and disposed, so it was reasonable to assume that offering to pick them up for free would be a long-term competitive advantage. Unfortunately, as demand for the waste by-product increased, companies started to sell the waste, dealing a severe blow to the financial and marketing model of the company I was working with.

4.1.2 Strategic Opportunism

At the opposite spectrum, we find a strategic philosophy called *strategic opportunism*. In this mindset, the company is constantly on the lookout for new opportunities in the market, making short-term adjustments to generate profits. As such, start-ups, with an entrepreneurial culture, are well adapted to seize these opportunities without creating too many disruptions in their company, creating new products and services on a more fluid basis, while less profitable activities can be de-emphasized, divested, or simply dropped. Of course, a close and constant appraisal of the market is necessary to identify and evaluate opportunities.

For example, a start-up I was working for was developing an application in digital health using artificial intelligence to evaluate self-care health opportunities. The first iterations had a web app and an online portal. As research indicated the use of the online portal would be minimal, development on it was dropped. As opportunities with a coaching component emerged, it was added to the business plan, and the overall marketing strategy was adjusted accordingly.

One of the main advantages is that strategic opportunities energize the organization and its team. They create opportunities for growth and an excitement for ownership to research and identify new opportunities. But the main problem is the advent of strategic drift, where a company is constantly making a series of short-term decisions which drive it away from its core mission and competencies. Also, short-term opportunities which could generate profits are rationalized as part of the opportunistic strategy, but can ultimately cost large amounts of resources for the organization to maintain. In the end, strategic drift could lead to a company being engaged in a series of markets or opportunities they did not have the internal resources and competencies to handle.

Going back to our earlier example, let us imagine that the start-up I worked with started getting multiple requests for "life coaching" services to supplement its health coaching. Seeing it as a small investment (in terms of time and resources to modify its platform), it accedes to it. It then identifies an opportunity for life and health coaching seminars. Having access to a database of health and life coaches, as well as potential clients, it goes forward with it. During one of these seminars, it is approached by a third party to build a portal of self-care health resources for individuals. From one small (profitable) decision to the next, our health app company has now become a company organizing coaching seminars as well as managing/offering support to an online portal for lifestyle and coaching resources.

4.1.3 Strategic Adaptability

This strategic philosophy, like the one before, is based on the assumption that the market is always changing, and companies have to be ready to respond to these changes. However, the difference with strategic opportunism revolves around the company being proactive to market changes, rather to reacting to these opportunities. As such, effort and resources are dedicated to understanding market trends, market opportunities, and potential shifts.

This philosophy implies putting into place tools that monitor and gather potential client feedback and comments. It also requires a culture that adapts to change. Shifting from one strategy to the next requires an open mindset, and changes can come quite rapidly as the information becomes available. But rather than aiming for short-term profits, a strategic vision has to be put into place to identify pilot projects and adapt, responding to expected trends. This is a philosophy where a company's culture must be very entrepreneur-focused.

This philosophy is nonetheless very risky, as companies have to correctly read incoming trends, lest they invest in the wrong trend and execute it incorrectly. As such, misreading an incoming trend could severely hurt and even end a small company. Also, beyond the risk of wasting resources, the company's brand and image could be irreversibly hurt, inhibiting future decision making.

4.2 Choosing Your Marketing Model— What Type of Company Are You?

The next step in building your marketing strategy is to define your marketing positioning, both for your company and your product. This means making a series of decisions on how your company will be oriented toward the market, how you expect to interact with your customers, and how you will be positioning your product. These decisions will be crucial going forward in selecting your marketing tools, your commercial channels, your distribution model, and more.

There are a number of marketing models a company can adopt, which we will be reviewing in detail in the following sections. These are the classical models (the production model, the product-focused model, the selling model, and the marketing model), which we will follow with a presentation of a few more modern concepts (relationship marketing, integrated marketing, and internal marketing).

4.2.1 *The Classical Company Models*

The first four models we will be examining are the classical company models: the production model, the product-focused model, the selling model, and the marketing model.

4.2.1.1 *The Production Model*

In this model, the company believes that its target consumers prefer to purchase widely available inexpensive products. As such, the company dedicates its effort in obtaining the best price for its products, focusing on lowering operating and material costs while increasing product volume and making sure products are available to as many people as possible in a consistent manner. It is one of the easiest concepts to implement, since it is focused on the mass production of goods. It could make sense to adopt this simple model when you have low fixed costs (location, equipment), low-variable costs (access to inexpensive labor and raw materials), or if you are facing a huge untapped market and well-defined demand. In this marketing model, the actual needs of the consumer are secondary compared to what the manufacturer is able to produce.

In life sciences, companies manufacturing generic pharmaceuticals and basic medical devices (basic bandages, gloves) with captive customers can use this model, focusing on price reduction and mass distribution rather than on personalized interactions with clients. In a market where there is a lot of competition, where prices are controlled, or where products can be up-branded, this model would be of limited use.

4.2.1.2 The Product-Focused Model

In this orientation, the company centers its marketing decisions and efforts directly on its product, as it believes that its target consumer will always purchase the superior product. These customers will remain loyal if they believe they are being well served, or if they have access to the best variety of products. Hence, the company focuses on creating and delivering superior quality, improving the product's performance and its innovative qualities to craft its marketing message.

When choosing this path, a company has to be careful to not overly rely on its product to generate sales; customers do not necessarily seek out and purchase the "best" product all the time, and choosing this path means investing extensively in the marketing effort. Effort must be made to ensure the price, advertising, and distribution chain is correctly set up. Also, the company has to remember that its products and options are at the heart of the consumers' purchasing process. Hence, continuous improvement (of the product, as well as sales message) hinges on the correct identification and manifestation of the best options that directly answer a consumer's needs.

4.2.1.3 The Selling Model

In this model, the main effort of the company is dedicated to selling the goods itself. Hence, the consumer is perceived as a passive actor, rather than an active consumer researching key products. This model is used by companies that believe the public is not necessarily aware of the product, its advantages, and its benefits. In this model, the risk shifts a bit as the focus of the company shifts away from its product, being put into its sales effort instead. This is most often used with products that consumers do not go out of their way to compare or purchase. Classical examples could be cemetery plots or life insurance. In life sciences, it could be vaccines or nutraceuticals with long-term benefits (hence, products that are proactive on health rather than reactive).

There is additional risk in using this model since it is often associated to negative selling models such as "high-pressure sales pitch" or "multilevel marketing" sales models, both of which have negative connotations.

4.2.1.4 The Marketing Model

In the marketing model, companies try to create value by creating a better value proposition than their competitors. This value proposition, quite simply, is how the product is positioned compared to the company's competitors, emphasizing its best attributes compared to its competitors. So, e.g., two similar drugs might have the same mechanism of action, but one product might emphasize its value proposition around its faster-acting mechanism, better impact on the digestive system, and better taste, while the second product could focus on the smaller size of the pill, its wider availability in drugstores, and the recommendations given by doctors. Both equivalent products in this case are emphasizing their unique value proposition to the customer, trying to get them to purchase their product.

In this case, the company is trying to convince the purchaser that its product is better than others because of the key attributes it has identified. If a customer perceives two products as equivalent, it will revert back to traditional decision factors such as pricing, availability, and quantity.

The key to a good marketing model is to be able to emphasize the customer needs and tie them back to your key attribute. If, e.g., you are targeting a population for which comfort is important (e.g., elderly people), then emphasizing the lack of secondary effects and the softer impact on the digestive system would make more sense. Another company, targeting individuals who need third-party validation during their decision-making process, could emphasize the doctor recommendations. A company focusing on the marketing model will have to be able to answer these four simple questions:

1. What market am I targeting?
2. What are their needs and demands?
3. Which key unique selling point will most likely impact their decision process?
4. How can I best deliver on this value proposition?

As an example, a few years ago, I was working with a client that had developed a family of skin care products for individuals going through radiotherapy. We did an online survey of potential customers to measure the key purchasing attributes of individuals using skin care products during radiotherapy. Some of the key attributes which consumers put emphasis on turned out to be around clinical validity and the ability to get the job done. As such, marketing collaterals were rewritten to emphasize the scientific origin of the product and its clinical validity (the products had gone through a series of clinical trials), and more space in the marketing literature was given to customer testimonials.

4.2.2 Modern Marketing Concepts

4.2.2.1 Relationship Marketing

Relationship marketing focuses on the creation of long-lasting relationships between consumers and the company producing the product in an effort to retain customer business. As such, the premise is that retaining a client is a lot less expensive than acquiring a new one. As such, the company focuses its efforts on creating strong connections with its clients by providing them with information and opportunities to interact with the company, as well as with other consumers. Successful relationship marketing companies usually have a greater understanding of their clients; they use tools such as Customer Relationship Manager (CRM) software to customize the customer experience and enhance the customer relationship.

4.2.2.2 Integrated Marketing

Integrated marketing is an approach that combine marketing actions into a single marketing activity which delivers superior values to each individual action. For example, a hospital purchasing expensive medical equipment would expect superior installation services, as well as comprehensive servicing of the machine, maintenance, and training on how to use the equipment. This concerted: effort would then result in a positive customer experience. The company's ability to offer all four services in a single point of contact demonstrates excellent integrated marketing strategy. When building an integrated marketing company, it is important to acknowledge that each marketing function is an important piece of the sale, and it is therefore crucial to coordinate how each of these fits with one another. The ability to coordinate these enables companies to distinguish themselves from their competitors.

4.2.2.3 Internal Marketing

Internal marketing makes an effective use of the company's internal resources for its marketing effort. As such, it recognizes that all company personnel must be customer-oriented. For example, personnel responsible for production must be committed to purchasing the best materials, accounting must be dedicated to offering a superior purchasing experience (invoicing the right amount, handling queries in a professional manner, responding within reasonable timelines, and so on. Furthermore, it recognizes that all employees can become spokespeople for the company, and that they must share a single positive vision of the product. To ensure this, companies are strongly encouraged to have a social media policy that aligns with the company's existing corporate culture and ethical guidelines providing the employees direction on how to interact when reaching out to public.

4.3 Creating, Adapting, and Implementing Strategy

As an innovative company, you are most likely developing new markets. But it is useful to reflect on what new business market you are creating. Fundamentally, you could be approaching the market through technological innovation, going from a component to systems strategy, targeting an unmet need, attacking a niche market, tackling a customer trend, or creating a dramatically lower price point. The next session will walk you through the different strategic decisions you will need to make.

4.3.1 Determining Your Target Market—Segmentation

We have already dealt in large part on the topic of the target market when discussing client analysis in Section 3.2.1. The important reminder here is that you can categorize your client by different segments, such as geographic, demographic, organizational, psychographic, and behavioral.

Geographic segmentation is choosing a client segment based on where the client is located. This could be as large as country, down to a few cities. For example, you might decide that you only want to target clients in a few states or provinces. The advantages to this are reduced regulatory complications (in life sciences, each country often has its own sets of regulations and laws) as well as reduced scope (advertising a product to clients who cannot purchase your products is a huge waste of money and effort).

If you are targeting individuals, then some form of **demographic segmentation** might be warranted. Demographic segmentation is selecting a client based on his demographic characteristics. It could be an age range, gender, religion, or more. The advantage with this type of segmentation is that homogenous groups within a same demographic segment often share many characteristics, including purchasing patterns. As such, a message written with the aim of targeting young adults will resonate much more than a broad message that does not talk to anyone.

Psychographic and behavioral segmentations are also useful when dealing with consumers' needs and requirements. Psychographic segmentation will be done when you target the lifestyles of your potential customers. Hence, you will be segmenting based on how a consumer's lifestyle, personality, and habits will dictate his purchases. Behavioral segmentation is done, keeping in mind, the needs of the consumer. The Kano model (Section 3.2.1.3) does a great job of demonstrating this.

Finally, if you are targeting organizations, then you will be most likely doing **organizational segmentation**. We have described this in detail in Section 3.2.1.2—the customer ecosystem- so remember if your business strategy is to interact with hospitals, GPOs, government agencies, commercial retailers, or any other organizations, your marketing strategy will have to reflect this as well.

4.3.2 Strategies Based on Products and Market

The first family of marketing strategies is based on Igor Ansoff's Product/Market matrix. Each of the four possible strategies is

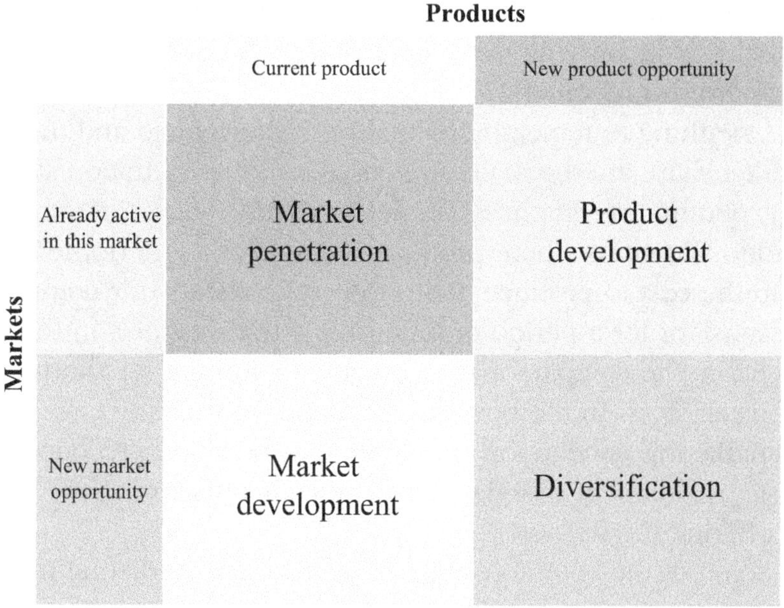

	Products	
Markets	Current product	New product opportunity
Already active in this market	**Market penetration**	**Product development**
New market opportunity	**Market development**	**Diversification**

Figure 4.1 Marketing strategies based on the product/market matrix.

classified based on the product and the current market presence. For the product axis, the company questions: Is this a product that I'm already selling, or is it a new opportunity? For the market axis, the company questions: Is this a new market, or is it the one that I'm already active in? This therefore splits into four strategies, namely, market penetration, market development, product development, and diversification (Figure 4.1).

4.3.2.1 Market Penetration

A common marketing strategy is **market penetration**, where a company targets a market in which it is already active with its existing products. The objective is to increase the sales of existing products, gain notoriety, and gain market shares. As such, success for this strategy is measured through increased sales volume, especially in relation to the overall estimated volume of the market. Market penetration can be attained by either convincing current clients to increase their purchasing velocity or identifying new client prospects (by taking them away from competitors).

One thing to remember is that both the volume and the market share are the main aims of a market penetration strategy, putting less emphasis on net profits. As such, this could lead to situations where products are sold at a loss (for less than the cost to produce them) to certain clients (for notoriety reasons), or for a period of time (like a trial session). In this situation, the company has to be able to account for short-term monetary loss. In the best-case scenario, a company engaging in marketing penetration should at least be able to recoup its costs by selling at cost, i.e., the cost to manufacturing the good or offering the service.

Some of the common ways of executing a marketing penetration strategy include the following:

- *Price adjustments*: This means lowering the price below that of the competitors, even selling at a loss, in the hope of increasing the sales volume. The objective could be two fold: increase the volume of sales to the point that the lower manufacturing cost makes the low price profitable, or increase prices gradually after the customers have adopted the product and the cost-to-change is too high. This is a very volatile way of entering the market, since it is very easy for your competitors to react.
- *Increasing visibility*: Additional promotions, visibility, and advertising of the product all generate additional visibility to current or unserved clients, generating more sales. The more people know about your product, the more they are likely to purchase it. This is especially true if you are expanding to new geographic markets or targeting new client segments. This can become very costly and should be monitored carefully (see Chapter 6 on metrics for more information on calculating the impact of your visibility).

4.3.2.2 Market Development

A company that has selected a **market development** strategy has decided to target new customers with existing products. It could be targeting new clients in an existing market, or it might have decided to expend in a completely new market (e.g., geographically or demographically).

Market development can be a tempting strategy, especially if one identifies a large untouched market. For example, companies will sometimes make broad statements for penetrating large geographic markets such as China, Europe, or the United States, without ascertaining if their product is a good fit for that market.

Also, a company can target **unmet needs** when they identify a specific issue a customer has. Unfortunately, the customer himself is often not a good source of knowledge for this type of information (he often has difficulty seeing beyond the traditional boundaries of existing products), so research on the client's current processes is necessary. Overall, you should be able to answer these four questions before going forward with a market development strategy.

1. Have I researched the chosen market that I wish to target to be able to understand it sufficiently?
2. How are the target client's base needs currently being met?
3. Are there any specific regulatory/legal constraints to targeting the chosen market, or the target market?
4. Will the product require considerable modification/adjustments before entering the targeted market?

There are two key approaches that are worth exploring for this type of strategy, namely, niche markets and price differentiation.

Niche markets are identified by fragmenting a part of an existing market into a specific submarket. This niche will logically be smaller than the large generic market, but also easier to reach and target by developing specific messages and using specialized channels to contact potential clients.

Niche markets will often be defined through demographic segmentation (age groups, revenue groups, gender), but also through factors such as lifestyles and personal preferences. A niche can also be developed by targeting a specific application.

A great example of niche marketing is one I developed with one of my clients, Nanopore Diagnostics (http://nanoporedx.com/). Nanopore had developed a rapid and portable platform for counting microbes right at the point of need. But it soon found that there were several companies addressing this market from a generic perspective, and it realized it quickly needed to find a specific niche if it wanted to distinguish itself from the pack.

As such, a wide scan of 20 different niches were examined to find those where the strengths of the technology (quick, portable diagnostics) would be invaluable. After literature reviews and interviews with a number of key opinion leaders (KOLs), an opportunity in salmon farming was identified, as this industry was in desperate need of a diagnostic kit they could use on-site to rapidly evaluate the presence of contaminants so that they could prevent the contamination from spreading rapidly throughout the farm. Currently, Nanopore is working with stakeholders and researchers to further adapt its product to this niche opportunity.

Another approach is to create different pricing policies to address different markets. This is called **price differentiation**. An example of price differentiation occurs when a product is available at full price for the regular public, and at a reduced price for students and seniors. The biggest challenge when modeling price differentiation is to establish barriers around the market so the consumers purchasing products at full price are not able to buy it at the discounted price. A way to address this concern is to develop a new different sub-brand so that the two brands do not cannibalize each other.

This strategy is used often by big pharmaceutical companies when selling their products in underdeveloped markets. As such, for these companies, price discrimination is an attempt not only to recoup research and development costs, but also to make drug prices sensitive to different degrees of price elasticity—responsiveness of consumers to changes in price [1].

4.3.2.3 Product Development Strategies

Companies that engage in **product development strategies** can offer new and improved products, as well as suggest new usage of its current products to their existing customers. These companies can identify **customer trends**, and so they can adapt and develop products that directly answer these needs. The challenge when addressing a new market is correctly addressing the price points at which customers will purchase the product without turning back to their regular products.

One of my clients utilizing a product development strategy around customer trends is JustBIO (http://english.justbio.ca/). JustBIO is a biotechnology company that has developed a series of health drinks, each adapted to a specific customer trend, rather than selling one generic health drink. Having developed a product designed to promote a global mind-body response by acting via the gut-brain axis through epithelial TLR2 (an evolutionary-conserved molecule that interacts with the gut microbiota), the company proceeded with the development of the JustADAPTÉ™ products line, which included products targeting trendy consumer concerns such as a natural solution to fight stress and anxiety, one to improve endurance and increase the level of energy, one to keep the cardiovascular system healthy, and one specifically designed for women's health. By developing a series of customized health drinks, JustBIO can distinguish itself quite favorably compared to more generic health drink brands.

A company that has developed a new **technological innovation** can focus on it to develop its new market. As we have discussed previously, it is crucial for the technology innovation to be related to the customer need. It should focus not on the technology aspect but on the client's needs, and how it is improving the consumer's life. Beyond new product innovation, two common strategies in this space are components to systems and product bundling.

Components to systems approach includes products that are designed when a company uses multiple components which were previously sold separately, and combines them into a single new product. A company using this approach changes the market significantly, as competitors can no longer compete to be the best component, but must rather find a way to redefine their offering, lest they be forgotten by consumers, who prefer integrated solutions.

The new generation of cold medications (like Tylenol Cold), which combine medication for headaches, nasal congestion, dry cough, fever, and aches and pains in a single tablet, is a great life science example of a "components to system" approach.

Product bundling occurs when a company bundles two or more products sold separately into a single new product offering. For example, a company could decide to bundle a cream product and a lotion-based product together in an effort to increase sales, while providing more value at a better price to customers. Research has shown that while this is often a successful strategy, it can only work if products are still available to be sold separately [2], as consumers dislike being forced into inflexible product solutions, and research has shown that people believe that if they put off their purchase, they will most likely be able to eventually buy the product separately.

4.3.2.4 Product Diversification

The fourth strategy is **product diversification**. This occurs when a company decides to develop new products for customers they are not currently serving. This is by far the riskiest of all four strategies, as the company creates risk on two distinct fronts: it is engaging in product development where it might not have the technical, marketing, and financial expertise it needs to ensure success, while simultaneously, targeting a market it has not previously engaged. This means it must acquire skills, facilities, technology, and expertise, thus increasing the complexity of the project [3]. Product diversification often results in alliances, mergers, or acquisitions of third-party technologies or entire companies.

Diversification can be either concentric (related to the company's current technology, product, or service) or conglomerate (totally unrelated to the current product portfolio). If a natural health company specializing in products for kids expands its product line to add a product for seniors, it would be going through concentric diversification, while a company that expanded from health products to bottled water would be engaging in conglomerate diversification. The first is usually done by a company that believes it can leverage its existing internal resources to extend to a new market without too much risk, while conglomerate diversification occurs when a company has identified a very profitable opportunity it believes it can successfully monetize.

A company's diversification strategy can be described by where it is moving in the value chain, be it horizontal or vertical.

A company engaging in vertical diversification would have identified a business opportunity outside its current business stage. For example, a company selling natural products, which were previously contract manufactured, decides to either develop internal capacity (its own production setup) or acquire external capacity (acquire its contract manufacturing partner) to move from a position where it is only selling, to a position where it is both producing and selling products.

Meanwhile, a company engaging in horizontal diversification would have identified a product opportunity in the same operational stage that it is active in. It could, e.g., go from selling a unique dental product to adding an ophthalmic product line. The objective in this case is to leverage the company's successful name and its brand as a successful distributor of high-quality products to different unrelated categories.

4.3.3 Marketing Strategies Based on Competitive Advantage

Another great author, Michael Porter, built a famous strategy framework, where the strategy is based on the scope and the source of the sustainable competitive advantage. The scope defines how you are approaching the market. It can either be narrow (a very focused slice of the market) or broad (a very large part of the market). As for your source of competitive advantage, it can either be around cost or differentiation (have a high-quality or unique product). Four strategies emerge from this grid: cost focus, cost leadership, differentiation, and differentiation focus (Figure 4.2).

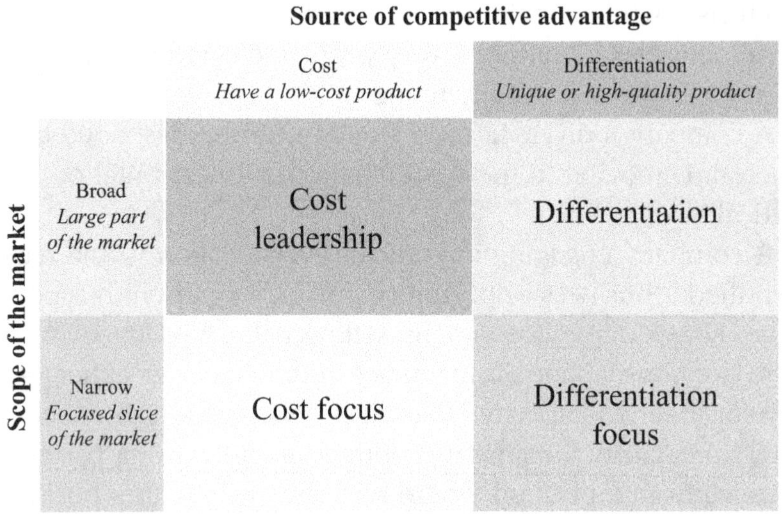

	Source of competitive advantage	
	Cost *Have a low-cost product*	Differentiation *Unique or high-quality product*
Broad *Large part of the market*	Cost leadership	Differentiation
Narrow *Focused slice of the market*	Cost focus	Differentiation focus

Scope of the market

Figure 4.2 Porter's four marketing strategies.

4.3.3.1 Cost Leadership

A company with a low-cost structure and a broad market may attempt to position itself through the introduction of a greatly reduced price in the market, initiating a **cost leadership strategy**. This is accomplished by continuous improvement of operational costs (paying employees less, saving on manufacturing costs, economies of scale, and buying in bulk). As such, the company strives to be among the lowest-priced products on the market. In this space, there are two sub-strategies which are low-end disruptive innovation and new-market innovative disruptor.

Companies use *low-end disruptive innovation* strategy when a simpler lower-cost product can be sold to disrupt an industry which has long focused on upselling better products for more money. The potential clients are those not interested in purchasing high-scale product, and can be targeted with these more accessible options. Usually, companies that are already selling in this market are not interested in selling to these clients as they are regarded as low-margin opportunities.

Alternatively, there are *new-market innovative disruptors* which target non-client entities, which are clients who initially had little to no interest in the products, balked at the more expensive existing options, and could be interested in purchasing a lower price point simple option. For example, a company that would have designed a simple yet inexpensive device to measure heart rate could compete with the traditional companies in this space, targeting clients who were interested in having their heart rate monitored, but not interested in paying full price for an expensive device.

While the two strategies focus on lower margin customers, the main difference between them is that low-end disruption focuses on overserved customers, while new-market disruption focuses on underserved customers.

4.3.3.2 Differentiation

Companies that have a sustainable competitive advantage and a broad market definition can use a **differentiation strategy** to focus on their unique dimensions that generate value. As such, since they have been able to identify something that they do better than everyone else, and are able to protect this competitive advantage, they can focus on differentiation to charge a higher price and generate higher margins per product. Some competitive advantages can include unique technology (e.g., protected by intellectual property or commercial secret) or a unique brand. This strategy is appropriate in situations where the client is able to afford the higher prices. It is possible to attain differentiation through a variety of tactics. Table 4.1 gives an executive example of this.

Table 4.1 Examples of Differentiation Tactics

If You Focus on the…	Examples of Differentiation Strategies
Product	New product improvement, adding new features, offering better quality, or offering an improved appearance
Price	Premium pricing, i.e., selling your product more expensively than your competitors
Brand	Developing a unique selling proposition or having a reputation for high quality and customer service
Promotional activity	Organizing intensive marketing activities and having high advertising spend
Distribution	Opening your own unique outlets, developing your own unique distribution network

Table 4.2 Summary of Marketing Strategies

Strategies Based on Products and Market	Objective	Typical Tactics
Market penetration	Increase sales of products, gain notoriety and market shares.	Price adjustments
Market development	Target new customers within existing market	Niche market
		Price differentiation
Product development strategies	Create new products for existing customers	Technological innovation
		Components to system
		Product bundling
Diversification	Add new product lines to target new market while leveraging current brand equity	Concentric
		Conglomerate
		Horizontal
		Vertical
Strategies Based on Competitive Advantage	*Objective*	*Typical Tactics*
Cost leadership	Offer a greatly reduced price compared to competitors	Low-end disruptive innovation
		New-market innovative disruptor
Differentiation	Focus on unique dimensions that generate value	Focus on product
		… on price
		… on brand
		… on promotional activity
		… on distribution
Cost and differentiation focus	Identify a sub-segment of your market to target exclusively	Niche marketing

4.3.3.3 Cost and Differentiation Focus

The cost focus strategy is established by selecting a key segment of your market and targeting it exclusively. These are groups with special needs, currently ignored by some of the larger competitors. As such, the idea is to identify a specific subset of customer, offering them something unique that other competitors cannot. Hence, by tailoring your marketing mix to these specific customers, you can uniquely address their legitimate requirements.

This is akin to the niche strategy we discussed earlier. It is also possible to attract customers through superior client relationships; the service to and interaction with customers in their market can build strong brand loyalty with them, making their market segments less attractive to competitors.

4.4 Developing the Marketing Mix

The marketing mix will be the set of tools and tactics you implement to operationalize your marketing strategy. To be useful, it will have to be as concrete and functional as possible.

We will be discussing the marketing mix through the lens of McCarthy's four Ps. If you do some research on the topic, you might find endless variations, with authors making great arguments about adding anywhere from 1 to 4 more "Ps." My perspective is that the four Ps still provide a strong backbone to building a quality marketing mix. When possible, we will be modernizing our discussion on each section, ensuring its validity. For example, while Porter might not have imagined the importance of digital ads and search engine optimization (SEO), we will be discussing both in promotion. The next few pages will be dedicated to the product, the price, the promotion, and the place.

4.4.1 Product

Your product (or service) has to fill a specific customer need (or target an unmet need). One of the ways to articulate this is to develop your unique selling proposition (USP).

Simply put, your USP is the characteristic (or set of characteristics) which only your product has, and that your competitors cannot and will not offer. It has to be compelling enough to entice customers to purchase your product, and avoid terms such as cheaper (cheaper is often associated with less value, which is contrary to what your USP is trying to drive) and better (which is a challenging and expensive claim to prove).

4.4.1.1 Key Product Decisions

There are four key decisions that you must make relative to your product:

1. **Quality decisions**: The quality of the product is directly related to your marketing strategy. As such, if you decided to sell an upscale product, investments in quality will be a must.
2. **Features decisions**: The optimal mix of features will have to be determined prior to commercialization. While there might be multiple options, you will have to shortlist the more promising ones. A good way to decide is to ask yourself: "Will this feature add to the perceived and actual benefits of the product?"
3. **Packaging decisions**: The packaging decisions you will have to make include the protective value of the packaging, visibility (especially if available in a retail setting), cost (some packaging can account for up to 40% of a product's total cost), and legal requirements.
4. **Branding decisions**: Your brand can generate sales as well increase confidence in the product's quality and

reliability. A solid brand image could be successfully used in future products as well.

4.4.1.2 Measuring Product-Market Fit

For start-ups, it is important to be able to quickly measure if your product is a good fit for the market. One way to do this is to measure product-market fit or PMF.

PMF is the validation that the value proposition in your product creates actual customer excitement to purchase your product, which will translate in actual sales. This follows the Problem-Solution Fit, which is the identification of the best solution to solve a problem. The difference between the two is quite noticeable: you might have developed a product that is the perfect solution to a problem, but which nobody will want to buy (due to cost, complexity, and so on). As such, measuring PMF ensures that you measure the market intent in purchasing your product.

The basic methodology to measure PMF revolves around engaging conversations with potential clients. As such, the entrepreneur contacts early adopters and evaluates their interest in the product or service. The principle is that by interacting with your early adopter you can identify something they react positively to, and then focus on your product or service on that key element. If after your investigation your current product does not have a market fit, your efforts will have to be dedicated to maximizing or retooling your product. This might mean changing the target market, changing the product, or changing both.

From a practical perspective, this PMF entails contacting potential clients directly. This is the only way you can develop a better understanding of their needs. You could contact them directly, interview them, meet them at conferences, and so on. Developing a deep understanding of potential clients is essential, in order to develop a significant value proposition.

So how do you know you have a good market fit? There is no exact number: more is better, obviously. But right now, many practitioners seem to be in agreement around the 40% rule. Hence, if at least 40% of the potential customers you talked to expressed strong interest in your product as something they absolutely need, then it is a good indicator you have a good market fit in your hand. If it is lower than 40%, you might struggle to gain significant traction to reach your market. You might need to redefine your target market, or you might have to redefine your product. Sometimes, the solutions might be highlighting the most compelling attributes, dropping some less popular attributes, or revisiting the product completely.

If you are already selling your product, some practical hints that you have not achieved PMF are that word-of-mouth marketing around your product has been ineffective, usage is not increasing significantly, media and press interest around your product is weak, the sales cycle is too long (and its growing longer), and many deals just never close. Inversely, if you are hiring sales and customer staff as fast as possible, sales are faster than your production, and reporters are calling to learn about your product, then you most likely have achieved market product fit.

One of the things to remember when checking for PMF is if you are evaluating the customer's interest in a solution that you are offering, or are merely investigating a problem/solution fit; if you are doing the latter, you might not necessarily be evaluating the customer's interest in the product or service you have developed, identifying a false positive, an interest in the market for a solution to a problem, but not necessarily that your product is the solution people are looking for.

4.4.1.3 Making Key Product Decision—The ICE Score

The ICE score is a simple method you can use to help prioritize your key product decisions. For each idea, you examine three simple aspects relative to your feature: its **impact**, its **confidence**, and its **ease**. For each aspect, you grade them on a scale of 1–10, averaging them out to calculate the ICE score. The higher the ICE score, the more priority the feature should be given.

The first question you ask yourself is "How much **impact** will the new product feature have?" A minor improvement, such as tweaking the menu configuration on a medical device, would generate only a small impact on the overall product value, while a large improvement, such as having better battery life on a portable device, would have a lot of impact performance.

The second question is "How **confident** are you that this feature will be useful?" An entrepreneur who already has people clamoring for a specific feature could easily claim a very high score, whereas a product idea that has yet to be market-tested would not have the same assurance.

The last question is "How **easy** will it be to bring the feature to market?" Alternatively, "How many resources are necessary to bring the feature to market?" If the feature is easy to implement and you already have the know-how, then a high number is expected. If the feature requires considerable resources with a lengthy testing cycle, a very low score is expected.

After reading through, you might be asking yourself: Isn't this scoring process too subjective? Why is something an 8 and not a 7? Wouldn't two people score them differently? Or even worse, wouldn't the same person rate the same product feature differently if he did the test twice? Couldn't someone manipulate scores to push through an idea?[*]

Just remember that the ICE score is not a system to prioritize ideas individually, but rather relatively to one another. The postulate is that as entrepreneurs we generate way too many ideas that we can act on, and we need a classification system to help us choose which idea to prioritize. So, while I might give

---

[*] Honestly, if you have this issue in your organization, you might have to reassess how ideas and collaboration occur. These problems clearly go beyond the ICE model.

Table 4.3   Example of an ICE Score Table

Feature	Impact	Confidence	Ease	Score
Portability	3	3	2	2.7
Higher and more accurate readability	8	6	4	6.0
Wi-Fi for results sharing	6	2	4	4.0
Lower stabilization time	6	8	8	7.3

a feature a 9 for impact, the score in itself has no value, other than being higher than another feature's 8.

To ensure comprehension, let us walk through a simple application of the ICE score. In this situation, a start-up has developed an innovative lab balance. With limited funds, it has decided to limit product development to only the most valuable features. After brainstorming, the team has whittled it down to four different feature possibilities: improving portability of the device, improving readability, including a Wi-Fi component to share the results in an app, and improving the device to have a lower stabilization time (Table 4.3).

Using our framework, we see than an idea having a lower impact (lower stabilization time) is most likely the feature to develop first since the team has more confidence in the impact it would have, and would be able to attain this objective more easily than the other options.

4.4.1.4 Building a Better Product—The Hook Model

Another model that can be used to optimize your project is called the Hook Model, which was developed by Nir Eyal [4]. It provides a framework for entrepreneurs on how to design products in sync with the way consumers purchase products. By focusing on how products can be habit forming, a company can design a product that integrates itself into a user routine, and build habit-forming products.

The objective behind the Hook Model is to form a relationship between a user's problem and a company's solution, reinforcing the relationship with each usage, with the intent of forming a habit. The Hook Model has four parts: a trigger, an action, a reward, and an investment (Figure 4.3).

The reward follows the action that was suggested by the trigger, in expectation of getting the reward. This reward in turn changes the behavior by gratifying the user, reinforcing the intent to repeat the action, and creating anticipation. Finally comes the investment, where the users invest themselves into the product or service, locking themselves in, simultaneously generating more value for themselves.

In healthcare, innovators that develop devices that directly target consumers can certainly learn from using this model. For example, wearable technologies (such as a FitBit watch) arguably follow this model. Upon successfully meeting the daily goals, the watch buzzes slightly (trigger). This causes the user to look at his watch, or even the phone app (action), to take a look at his successful goals (reward). The app also stores and analyzes data over time (investment), hence storing value so the user can benchmark himself. The user finds himself locked in because if he switches devices, he will not have access to this stored value anymore, hence not being tempted to move to a competing product, unless this product offers to import existing data.

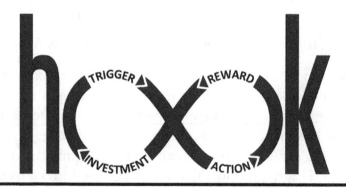

Figure 4.3 The Hook Model [4].

4.4.2 Pricing Strategy

Pricing is the amount the customer is willing to pay for your product. It is often a combination of perceived and relative value. Of course, pricing is also directly tied to your marketing strategy. As we discussed earlier, if your marketing strategy was market penetration, your pricing strategy should directly reflect this decision. When developing your pricing strategy, the factors to consider include the product itself (cost to manufacture), perishability, distinctiveness, and its space in the life cycle.

The basic pricing strategies are cost-based, competitive-based, or customer value-based.

4.4.2.1 Cost-Based Pricing

The idea behind cost-based pricing is that the sale of each product should generate value (i.e., profit) for the company. As such, these companies base their overall price on the cost of the product, and by adding a percentage to that cost as a profit margin.

Companies will have to determine the base cost of their product to properly determine their price, which would be based on their fixed costs (costs that are independent of volume sold) and their variable costs (costs that vary according to total volume sold).

The main challenge with this type of pricing is justifying it to the customer, who might not perceive the same value around the product. For example, if you have an expensive manufacturing process and your product cost is quite high, even adding a small premium might take the product out of the acceptable range of what the client is willing to pay.

4.4.2.2 Competitive-Based Pricing

Competitive pricing is based on conducting an overview of the competitors' current pricing, and then adjusting your price relative to your strategy. If your strategy was market penetration, then you would set the price lower (hence engaging in penetration pricing). If you had a strategy of pricing distinction, then you would try to price it higher (engaging in premium pricing).

One of the issues with fighting on pricing is that *"Wars begin when you will, but they do not end when you please* (Machiavelli)." Since competitors are free to react to your pricing, it will most likely lead to a price war. Some ways to avoid a price war can include revealing your strategic intentions, competing on quality, or co-opting contributors. If you are already engaged in a price war, some of the ways to respond include engaging in complex pricing options (bundling, quantity discounts, and loyalty programs) or introduce flanking products [5].

4.4.2.3 Customer Value-Based Pricing

Customer value-based pricing prices the product based on what the customer is willing to pay. This may require a market test or a customer survey to properly determine the price. It is quite challenging to find a price point that would be satisfactory to both you and the customer, since the client has a tendency to undervalue products and will generally prefer lower priced goods. This can lead to undervaluing your products since customers seldom value products to their full value. Possible reference points for consumer value prices include last paid price, historical competitor prices, expected future price, and usual discounted price.

4.4.2.4 Price Skimming

Price skimming occurs at two points in time. At first, pricing for the product is quite elevated, so that early adopters who really want your product (and who are a lot less price-sensitive) will pay top dollar for it. As time progresses, the price of the product is lowered, increasing sales velocity, until the company reaches its target price. Skimming enables the company to recoup costs quickly, as the few sales that are generated at first are compensated by higher price per product.

4.4.2.5 Freemium Pricing

Freemium pricing (a combination of the words free and premium) is used mostly with digital products although the main product is offered for free, the customer has to pay a premium to access advanced features or services. The basis of this model is that the freemium product is fully functional by itself, and the options that must be purchased only serve to enhance the program, and by offering a free version, designers expect rapid product adoption.

For example, Epocrates (http://www.epocrates.com/) is a medical reference app used by thousands of doctors. The free version can be used to review drug prescribing and safety information for thousands of brands, and generic and over-the-counter (OTC) drugs. Drug information includes adult and pediatric dosing for FDA-approved and off-label indications and black box warnings, contraindications, adverse reactions, and drug interactions. The pay-for-use version can give extra information such as disease information, infectious disease treatment, and lab and diagnostic information.

4.4.2.6 Other Considerations for Your Pricing

4.4.2.6.1 Price Discounts and Allowances

There are a number of situations in which a company will adjust their official prices and give discounts. Some examples include the following:

- **Early payment discount**: Many companies will give a price discount to companies who pay on time. For example, a company might give a "2/10, net 30" discount,

which would be 2% discount on payments that are made within 10 days of delivery, or full payment in 30 days.
- **Quantity discount**: This is given to companies that reach a predetermined volume of purchases. For example, a company could have a pricing policy such as "$10 per unit if you purchase less than 500 units, $9 per unit if the order is over 500 units." These are usually offered to all customers equally.
- **Allowance**: In some situations, some companies will give a specific allowance to encourage sales. This can be trade-in allowances (granted when returning an old product and trading it for a new one) or promotional allowances (for participating in promotional programs).

4.4.2.6.2 Price Endings

An old pricing trick is to end the price of a product with a 9 rather than a 0: as such, a price of $299 is viewed as much less than $300. This is because of our tendency to read from left to right, rather than rounding, which leads us to judge by the first digit, an effect that is also called the drop-off effect [6]. This makes the $300 product a lot more expensive in a customer's estimation. This in turn has led to prices ending in 9 being perceived as bargains by customers. In one study, a product's price was raised from $34 to $39, and the researchers observed noticeable increase in sales, where the prices ending in 9 was able to outperform the lower prices on average by over 24% [7].

4.4.2.6.3 Pricing in Pharmaceutical Products

If you are developing your pricing strategy for a pharmaceutical or biotechnology product, you might face a number of specific challenges such as the following:

- Insurance agencies/government reimbursement policies: In many locations, agencies fix a set price at which drugs and devices will be reimbursed. If a product's price

exceeds this list price, the remaining cost must be borne by the consumer directly; these pricing policies have been criticized in some countries for being non adaptive to high-price biologics, thus triggering medicine shortages as production for these drugs is scaled down [8].
- Media and consumer group backlash: Due to recent issues around pharmaceutical pricing, the media and consumer groups are especially wary of high prices in pharmaceutical and biotech products. As such, products with exceedingly high prices are subject to monitoring and even negative media campaigns.
- Company self-regulation of prices: In many states and countries, the company's ability to self-regulate prices for drugs and medication is being put into question. Recent headlines such as "Some prescription drug prices rise by 1,000% in 2017" [9] or "Cancer Drug Price Rises 1,400% With No Generic to Challenge It" [10] have only added to the controversy. Some CEOs in big pharmaceutical companies have come out recently advocating for more industry action relative to concerted drug pricing.

4.4.2.6.4 The Role of Group Purchasing Organizations and Purchase Orders on Pricing

If your goal is to sell your products or services to hospitals, you might find that the organization is a member of a larger group purchasing organization (GPO). GPOs are (usually) not-for-profit organizations that aggregate the purchases of its members (typically hospitals, nursing homes, and healthcare organizations); as such, they leverage the potential purchasing volume to negotiate lower prices from vendors and distributors.

GPOs are very frequent intermediaries when selling to healthcare organizations, accounting to up of 72% of all hospital purchases in the United States [11]. When dealing with them, you will find that the process is much more formalized and includes various processes such as Request for Proposals (RFPs) and Requests for Quotations (RFQs). Obligations to purchase within GPOs will vary from one contract to the next. In some situations, organizations will be allowed to deviate from contracted suppliers, while in others members must purchase from the selected vendors.

4.4.3 Promotion Strategy

The promotional aspect in your marketing strategy will focus on how you communicate toward your potential clients. The objective of your promotion strategy will include generating sales, increasing awareness, and improving your market presence. This will be achieved using tools such as advertising, sales promotions, public relations, and direct marketing.

4.4.3.1 Objectives of Your Promotion Strategy

Your promotion strategy will usually focus on either short-term or longer-term goals. Short-term goals can include increasing sales, entering a new market, and introducing new offerings, while longer-term goals can include building brand awareness and increasing product recognition.

When the company is not really well known by its potential customers, the promotion strategy should be built around a more long-term perspective. During those campaigns, the company strives to increase the number of people recognizing the company, thus increasing the opportunities when the product will be purchased. As such, the objective is to increase recognition compared to other competitors through increased brand awareness (recognition of the overall company brand) or product recognition (knowledge of the product and its differentiating features). Brand awareness and product recognition usually increase over time through more long-term tactics such as public relations and recurring advertising.

When a company's objective is shorter term, it tends to favor tactics that have a direct impact on its sales. It is trying to generate new sales in existing markets, entering new markets or introducing new products, for example. As such, it will put forward a series of short-term activities that focus on increasing the sales of its product and services. It will engage in tactics such as advertising, sales promotions, and direct sales, using a message that matches the product's chosen strategy.

As you select your promotional strategies, you should make sure the results are measurable. For example, a company that decides to increase customer awareness could postulate that "The objective of the promotional strategy is to increase web visits to the web site by 400% over the next month" or "The objective is to double the number of requests for information over the next 3 months." We will be dedicating more space to metrics in Chapter 6.

4.4.3.2 Promotional Tools

There are many promotional tools you can use to reach your promotional objectives. Over the next few pages we will be going over a few, and matching them to short- and long-term objectives.

- **Advertising** includes most mass media tools that have extensive reach but little personalization. These include billboards, magazines, television, and radio. Of course, the more specialized the media, the more personalization you can include in your message. Advertising is usually used in a short-term perspective (generating sales), although if it is used consistently, it can be used to attain more long-term objectives around brand awareness. Advertising in life sciences can be subject to specific regulations: this is addressed specifically in Section 7.1, where we discuss "marketing and life sciences" at length.
- **Sales promotions** are initiatives that target consumers directly by giving them access to an immediate and direct benefit. The aim is to increase sales rapidly by directly impacting product price and motivating the client to

purchase. Sales promotions can include coupons, samples, referral programs, and loyalty incentives.

Sampling is a very popular sales promotion tactic used by companies in the pharmaceutical sector. As such, doctors are given free samples of products they can share with patients, or for their own use. The use of sample has been shown to influence doctor's prescription habits [12], and has the additional benefit for the patient of having access to medication quickly (not having to go to the pharmacy for his first dose).

- **Personal selling** is the hiring and coordination of individuals to sell the product in a face-to-face setting. This can include hiring sales staff, attending trade shows and conferences (where you can meet your clients), and having a dedicated showroom (where clients can come to visit and see your product). The advantage of personal selling is that it creates conversations with clients, and can be used to generate short-term sales (sales staff and showroom) or more long-term relationships (trade shows and conferences).

  Representatives in life sciences are particularly popular, and heavily used by companies in pharmaceuticals, biotech, and even medical technologies.

- **Public relations** (PR) is the leveraging of media to share a message with key decision-makers. PR documents can include articles in trade magazines, press releases, news media events/press conferences, and more. The difference between public relations and advertising is that advertising is a paid media where you maintain complete control over the message, while PR usually includes a third party reporting on your product, implying loss of control over the message.

- **Digital marketing** is at the confluence of the four previous categories and includes elements such as social media, blogs, YouTube, and corporate websites. We will be going over digital marketing specifically in Section 4.5.

4.4.3.3 Choosing Your Promotional Message and Channel

As important as choosing your promotional objective and the promotional tools is the choice of the promotional message. The promotional message will be what you want to say about your product and service, and what you want the customer to capture relevant to this message. As a general guideline to preparing this message, here are a few questions you can ask yourself:

- *Who will my message be addressed to?* By now, your target client must be very well defined. Nonetheless, it is crucial to keep that segment in mind, lest you start preparing a much more generic message that has little impact on your potential buyers. Keep in mind the product benefits and why these benefits are important to clients to ensure an optimized promotional message.
- *What differentiates me from other companies out there?* You can focus on your USP as part of your marketing campaign to generate interest in your product.
- *Which promotional tool is best adapted to carry my promotional message?* Figure out how your target client will best respond to your message. Does he need to hear it? Does he need a lot of information or short bursts of information so he can track and compare? Does he need to interact and ask questions throughout his purchasing process, or only be receptive to information?
- *What are my competitors doing?* Examine how your competition is already talking with your client. Is he giving samples? Is he engaging with them in direct selling, or is he running an advertising campaign? While it is important to distinguish your unique promotional campaign, it is also important to know what your client is already exposed to.
- Building customer profiles (as we saw earlier in Section 3.2.1.5) will help with how you build and address the

message, and will be invaluable in preparing your customer message.

4.4.3.4 Choosing Your Promotional Tools: The Bullseye Framework

The Traction Bullseye framework was conceived by Gabriel Weinberg and Justin Mares [13]. It is an effective process that a start-up can use to select the best promotional channels. It breaks down the process into five simple steps: brainstorm, ranking and classification, establish priorities, test, and focus (Figure 4.4).

The first step is to **brainstorm**, using the 19 promotional channels that the authors identified in their book as a framework. For each channel, try to reflect on how you could potentially use each, trying to match your strengths and opportunities to each of them. Try to evaluate them according to opportunities, cost, and potential strategy. Without going extensively over each channel, we have included an executive version of the 19 channels to help guide you in your first brainstorming steps (Table 4.4).

This is followed by the **ranking and classification** of the channels into three distinct categories. These are the inner circle (which channels are the most promising?), the potential

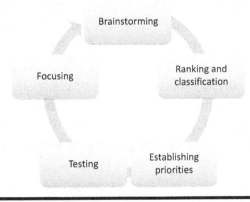

Figure 4.4 **The Bullseye Framework.**

Table 4.4 The 19 Communication Channels Used in the Bullseye Framework

Promotional Channel	Application in the Bullseye Framework
Affiliate programs	Paying a third party to generate leads of a sale
Blogs	Targeting the blogs that your potential clients read
Business development	Building strategic partnerships
Community building	Fostering community growth in your end users
Content marketing	Creating content to stimulate interest in your product without promoting a specific brand
Email marketing	Emailing messages directly to users
Engineering as marketing	Utilizing internal engineering resources to build tools that reach potential customers
Existing platforms	Leveraging existing platform (e.g. Amazon) with existing customer bases
Offline ads	Advertising in traditional media such as TV, radio, and magazines
Offline event	Sponsoring an offline event
Public relations	Pitching your story to traditional media outlets such as magazines and TV
Sales	Focusing on the action of selling the product
Search engine marketing	Purchasing advertising on web search engines
Search engine optimization	Improving your website ranking in search engines
Social/display ads	Purchasing ads displayed on social media websites
Speaking engagements	Identifying events where you can present

(Continued)

Table 4.4 (*Continued*)   The 19 Communication Channels Used in the Bullseye Framework

Promotional Channel	Application in the Bullseye Framework
Unconventional PR	Arranging anything from publicity stunts to unique customer appreciation initiatives (e.g.: signed letter from the CEO)
Viral marketing	Encouraging growth through current customer referencing

candidates (who could potentially work), and long shots (those that need a lot of things to work out right to be effective).

The next step is to establish **priorities** and select the top three prospects from your inner circle.

Once these top three channels are chosen, you are ready to go into the application. While we will be spending a lot more time on the implementation and establishing control mechanisms in Chapter 5, the main goal of the fourth step is to start by doing a small series of **tests** before investing fully into a channel. If you have selected trade shows, start by going to one, and then evaluate your returns before investing into a year-long road trip.

Finally, if all has gone well, one of your three channels will have given you better results. It is quite normal for one channel to dominate in the early stages of a start-up. Continue to **focus** on this channel. You can also restart the cycle, brainstorming on the 18 other channels, using the information you learned along the way to start developing new channels.

4.4.4 Distribution Strategy

Distribution and logistics seldom warrant much attention from decision makers: while efforts are poured into branding and promotion decisions, distribution is often an afterthought or implemented without much strategic thought. Yet, choosing the best distribution model and working toward optimization can generate close to 15%–20% of savings on the total product costs in life sciences [14].

As the name implies, the distribution strategy is the decisions you will make on how you will be distributing your goods to your end users and clients. A good distribution strategy is essential to your sales: if your clients cannot get their hands on your products, they will not be able to purchase it. For this book, we will be discussing decision factors that play into choosing your distribution strategy, followed by a brief description of different distribution channels that are available.

4.4.4.1 Factors to Choose Your Distribution Strategy

Many factors will influence the type of distribution strategy you select. We will be going through the main ones, namely, branding/positioning, your clients, and your competitors.

First, how you **positioned and branded** your product will have a huge influence on which distribution channels are available to you. For example, if you have positioned your product as a luxury/exclusive product, your distribution channels will include high-end and personalized selling methods which are better adapted to your product. If you have decided to position your product with a mass marketing appeal position ("A widget in every home"), large-scale distribution chains will be more your speed. Your distribution channel will also influence your brand presence and visibility, so it is important to take that decision in symbiosis.

Second, how your **clients** usually evaluate and purchase your product (or derivatives) has a huge influence on which distribution channels you select. Some of the considerations you will need to make include the following:

a. How do clients usually purchase this product?
b. Do they need any personalized interaction and education to use this product?

c. Does the product need customization once it is purchased?
d. Is servicing an important component of your product/service?

Once you have properly evaluated your client's needs and habits, match them to a relevant distribution strategy. For example, if your product requires little to no interaction for usage, you might focus on channels such as web-based or distributor networks. If your product requires heavy customization, you might need to include points of service and sales with specialized staff in your distribution strategy.

Also, consider how your **competitors** are currently selling their products. Since your competitors are already active and selling, this means that they have already generated client streams, which you will have to compete with as well. You can do this by taking advantage of any specific advertising strategy they have developed. For example, if they have an important web presence, you might need to develop your own web strategy as well.

Finally, if you choose multiple distribution channels, you have to be careful to minimize pricing conflicts between them. If you are undercutting your distributors with your own direct sales efforts, they will not feel as motivated to develop your product. Remember that consumers will always look for the lowest price points when looking at an equivalent/identical product, and it is important to map out your pricing strategy across distributors.

4.4.4.2 Different Types of Distributions

Distribution channels break down into two main families: store retailing (physical and ecommerce) and non-retailing (direct sales [Internet, catalog, sales team], wholesaler/distributors, value-added reseller [VAR], and sales agent/manufacturer's representatives).

Retail sales are the sales of finished goods to consumers or businesses driven by a retail entity. These occur at the end of the supply chain, and preclude further transformation by third parties. Retail sales can be done at brick and mortar stores, where the product is available physically, or through ecommerce channels (such as Amazon, Ali Baba, or eBay). As of 2018, ecommerce and digital goods sold through retail accounted for 8.8% of all retail sales. Countries such as the United Kingdom and China were some of the heaviest users of digital ecommerce [15].

There are a number of advantages of doing retail sales. First, by selling goods in a retail outlet, your goods will have an increased reach that you would have had going at it alone, as you will profit by the existing visibility of the retailer you are selling at. Second, you will profit from the existing customers visiting the retail outfits. Third, investments costs are considerably lower than building your own structure from the ground up. Finally, selling through the retailer puts the responsibility of selling the product on the retailer, rather than on yourself. In doing so, you can rely on his existing sales force, brand image, expertise, and existing clientele.

Non-retailing distribution channels include both direct and indirect sales. Direct sales are the initiatives where you are directly responsible for the relationship with the clients (Internet sales, sales team), while indirect sales are those where you rely on third parties to interact with clients outside of retail channels (such as wholesaler/distributors, VAR, and sales agent/manufacturer's representatives).

Direct sales can be done by setting up your own sales staff to meet clients on a one-to-one basis, by setting up a website to sell products directly to clients, or by printing and distributing a catalogue of your product. The main advantage here is that you maintain control over the sales process. You are responsible for your company's brand and as such, it is easier to manage how the product is sold. Feedback from clients also comes back to you much more quickly as you are directly interacting with them. The main disadvantage is the cost necessary to set up, maintain, and promote this type of initiative.

Indirect sales are strategies where third parties distribute your goods either through retailers, wholesalers, smaller entities/independents (sales agents), or by adding new products and features to an existing product to repackage and sell (VAR).

Wholesalers and distributers are intermediaries between the manufacturer and the retail outlet which specialize in the movement and sales of goods in specific industries. They usually have large specialized networks and staff on hand, making them experts in moving products. One of the main issues with these players is that your product can be one of many, and you might feel the intermediary is not doing everything within its power to move and distribute your product. Also, their intervention in the product cycle usually means that they will increase prices significantly.

Sales agents are usually small companies or individuals which sell your product on a commission basis. Having a smaller product portfolio usually leads to more personalized customer experiences and more attention to your product, but this can also lead to situations where they have limited reach to develop bigger clients. Sales agents can be used when developing a new market, with the intent of developing your own sales force later when the market is more profitable.

The last indirect alternative is VAR. These companies typically purchase products from other companies which they repackage as their own. As an example, imagine you have developed a fitness app. An authorized VAR could use your app and sell it as part of its health coaching services. The advantage is that it generates autonomous sales, but if your product is one of many on the resellers menu, it might not get the full focus it deserves. Table 4.5 summarizes well the major distribution strategies.

Table 4.5 Summary of Distribution Strategies

Distribution Strategy	Advantages and Disadvantages
Store retailing (Products are sold to a retailer, which in turns sells to consumer)	• Advantages: Increased reach, retailers have an existing customer base, selling is retailer's responsibility • Disadvantages: Identifying the best-suited retailer, retailer's cut can be high, loss of control over branding message
Direct—Internet (Setting up an Internet presence to generate sales)	• Advantages: Direct contact with client, control over message, higher profit margin • Disadvantages: Responsible for driving traffic to website, maintenance website costs, considerable initial investment
Direct—Sales staff (Setting up your own sales staff to generate sales)	• Advantages: Direct contact with client, control over message and staff, staff dedicated exclusively to your product • Disadvantages: Slow initial sales growth, considerable recurrent costs, important initial investment
Indirect—Wholesaler/distributer (Working with an existing company to distribute product to retailers)	• Advantages: Takes advantage of existing network and existing sales forces/market knowledge, third parties can hold an inventory product • Disadvantages: High mark-up prices, your product is one of many, and lack of accountability
Indirect—VAR (Companies package your product with their own)	• Advantages: Strong buy-in into your product, market specialists, and mostly focus on individual sales/small and medium enterprises (SMEs) • Disadvantages: Lack of reach for bigger accounts, product dilution can mean lack of focus and more expensive
Indirect—Sales agents (Working with independent agents who sell your product)	• Advantages: Low costs (commission-only model), you retain control of pricing and branding, more transparent than distributor in share market data • Disadvantages: Not a lot of reach, not always specialized, low buy-in into company means potential turnover

4.4.4.3 Specific Considerations for Distribution in Life Sciences

There are a number of issues that are relevant to distribution in life sciences, but we will be focusing on the issues relevant to the ecommerce, the cold chain, and the opportunities in regional consolidation.

Growth in ecommerce has changed the shape of retail industry. While ecommerce is only expected to account for 9% of total retail sales in 2018, this number could reach 20% by 2020 [15]. This means firms will be increasingly solicited to interact with clients abroad, in locations where your product might not be cleared for distribution, or in regions where specialized shipping and distribution is an issue. Integrating international ecommerce in your distribution strategy might become a necessary operation.

Managing the cold chain integrity is a complex issue. If your product is particularly sensible, you will have to spend considerable effort monitoring the product, and choosing your monitoring system of choice (e.g., choosing an active versus passive monitoring system). The last few years have seen increased consolidation in the market, a trend that might continue onward.

Also, it is important to keep an eye toward consolidation and how you can generate cost savings when intermediaries consolidate. While traditional distribution models had independent country structures for each country where products were available, the emergence of regional players has to stay within companies' mindset in an effort to reduce complexity costs.

4.5 The Role of Digital Marketing

Sidestepping the role of digital marketing is impossible at this point. Indeed, a recent study [16] found that the greatest growth for the pharmaceutical, diagnostics, biotech, and medical device marketing budgets is taking place in social media, mobile/tablet apps, and digital sales materials. For consumers, the greatest growth is taking place in mobile/tablet apps, social media, and digital ads. Hospital providers are also shifting to digital channels for marketing to healthcare consumers. Consider the following facts for "Google's The Digital Journey to Wellness: Hospital Selection" [17]:

- Seventy-seven percent of patients use search engines prior to booking appointments.
- Eighty-four percent visited hospital sites, and 54% visited health insurance company sites.
- Search drives nearly three times as many visitors to hospital sites, compared to visitors from other referral sites.
- Forty-four percent of patients who research hospitals on a mobile device schedule an appointment.

The digital marketing process is usually defined in two steps: awareness and conversion. Awareness includes all the activities in which the company is focusing on increasing its visibility and overall presence. This is followed by the conversion of visitor, from interested to potential leads. Converting website traffic into business leads or sales is the primary purpose for your digital marketing campaign. For this section, we will tackle two specific topics: driving traffic and selling products online. In Chapter 6, we will be going over the different metrics you can use to measure the efficiency of your digital marketing activities.

4.5.1 Driving Traffic

In order to generate sales, you must first ensure that consumers are able to reach your digital assets. There are many ways to do so, so here are some of the easier ones to put into place.

- **Generate quality content**: Write and distribute content such as press releases and articles on recognized media. You can do this in traditional media such as trade magazines, or more modern platforms such as LinkedIn. When you do publish, make sure to insert links into the articles that point to your corporate website in order to drive traffic to your digital assets.
- **Keyword strategy**: Insert related keywords into the content you generate. This will ensure your content and website show up more often in search results, generating more web traffic.
- **Website optimization**: Optimize your website on both traditional and mobile media. Having a clunky website is like having a clunky brochure, and generates high bounce rates.
- **Social media**: Post engaging social media posts to generate more traffic to your website. Use pictures and videos to make your content more engaging.

4.5.2 Selling Products Online

Here are a few elements to remember if you do decide to sell products online directly.

First, the ordering and shopping process should be simple for the customer. Browsing, finding, selecting, and purchasing should all be done without issue for your customer. If the consumer must struggle through a complicated purchasing process before having access to this product, he will most likely drop his purchase and move on. This also includes the transactional components (such as the payment platform), so make sure your transaction as smooth as possible.

Second, the information should be displayed quickly. It could be tempting to save money by using a service provider that offers "ok" service, but most customers will leave a website or will uninstall their app if they wait more than 8 seconds.

Finally, make sure that delivery of the product is on time, as this is a major element of repurchasing and customer fidelity.

4.5.3 Digital Marketing in Action—The Pirate Metrics: "AARRR!"

Dave McClure, former CEO of the 500 Startups venture capital fund, developed what is referred to as the Pirate Metrics to guide start-ups when acquiring and converting customers in digital marketing situations [18].

Reasonably easy to use, it intuitively breaks down the five behaviors of customers, describing them from the moment they learn of your company or product to when they spread the word. The "AARRR!" metric stands for Acquisition, Activation, Retention, Revenue, and Referrals (Table 4.6).

Table 4.6 Application of the AARRR! Framework

AARRR!		Description
A	Acquisition	How does your customer learn of your product? How would they come in contact with your company? Answering these questions helps you define the channels that bring the customer to you.
A	Activation	Did the customers do anything once they learned of your company? Did they sign up for a newsletter, activate an online profile? Did they get in touch with you? The objective here is that you want the customer to start interacting with your company, even if it is not yet a full purchase.

(Continued)

Table 4.6 (*Continued*) Application of the AARRR! Framework

AARRR!		Description
R	Retention	Did the user come back? How often does he come back? At this point, you will have to define what an inactive user is relative to your product. It could be he did not log on for a few months, or he did not make any purchases in the last quarter. Calculate and monitor this metric to make sure you are still hitting targets.
R	Revenue	How do you generate revenue with your users? Do they purchase products, bundling, or services? The final goal of any for-profit company is to generate revenue, so it is important to understand and share your revenue model.
R	Referral	Are you users spreading the word about your product? If not, how can you encourage referrals? If your users are bringing new users, you are lowering your customer acquisition costs.

References

1. Bhanji, S. Price 2012. Discrimination in Pharmaceutical Companies: The Method to the "Madness." *Harvard College Global Health Review.* https://www.hcs.harvard.edu/hghr/online/price-discrimination-method/ (Accessed January 29, 2018).
2. Derdenger, Timothy, & Kumar, Vineet. 2012. The Dynamic Effects of Bundling as a Product Strategy. *Harvard Business School.* https://hbswk.hbs.edu/item/the-dynamic-effects-of-bundling-as-a-product-strategy (Accessed January 30, 2018).
3. Ansoff, Harry Igor., 1968. *Corporate Strategy*, Penguin Books, Great Britain.
4. Eyal, Nir. 2014. *Hooked: How to Build Habit-Forming Products*, Portfolio, Toronto, Canada.

5. Rao, Akshay, et al. 2000. How to Fight a Price War, *Harvard Business Review*. https://hbr.org/2000/03/how-to-fight-a-price-war (Accessed January 30, 2018).
6. Bizer, George, & Schindler, Robert. 2005. Direct Evidence of Ending-Digit Drop-Off in Price Information Processing. *Psychology and Marketing* 22(10): 771–783. doi:10.1002/mar.20084. https://doi.org/10.1002%2Fmar.20084 (Accessed January 9, 2018).
7. Anderson, Eric., & Simester, Duncan. 2003. Effects of $9 Price Endings on Retail Sales: Evidence from Field Experiments. *Quantitative Marketing and Economics* 1: 93. https://doi.org/10.1023/A:1023581927405 (Accessed January 30, 2018).
8. Vogler, Sabine, et al. 2015. Pharmaceutical Pricing and Reimbursement Policies: Challenges beyond the Financial Crisis, Vienna, Austria. https://www.researchgate.net/publication/282619555_Challenges_and_opportunities_for_pharmaceutical_pricing_and_reimbursement_policies (Accessed January 9, 2018)
9. Heaney, Paul, & Hugues, Gwyndaf. 2017 Some Prescription Drug Prices Rise by 1,000% in 2017. *BBC News*. http://www.bbc.com/news/uk-wales-42506736 (Accessed January 9, 2018).
10. Loftus, Peter. 2017. Cancer Drug Price Rises 1,400% with No Generic to Challenge It. *Wall Street Journal*, December 25, 2017, https://www.wsj.com/articles/cancer-drug-price-rises-1400-with-no-generic-to-challenge-it-1514203201 (Accessed January 8, 2018).
11. Healthcare Supply Chain Association. Frequently Asked Questions, http://www.supplychainassociation.org/?page=faq, (Accessed January 8, 2018).
12. Sufrin, Carolyn., & Ross, Joseph. 2008. Pharmaceutical Industry Marketing: Understanding Its Impact on Women's Health. *Obstetrical & Gynecological Survey, Netherlands* 63(9): 585–596. doi:10.1097/OGX.0b013e31817f1585. PMID 18713478 (Accessed January 30, 2018)
13. Weinberg, Gabriel, & Mares, Justin. 2015. *Traction: How Any Startup Can Achieve Explosive Customer Growth*, Portfolio, Toronto, Canada.
14. Deloite. 2016. Innovative Routes to Market Rethinking the Life Sciences Distribution Model. https://www2.deloitte.com/content/dam/Deloitte/uk/Documents/life-sciences-health-care/deloitte-uk-lshc-innovation-routes-to-market.pdf (Accessed January 9, 2018).

15. Invesp, Global Online Retail Spending—Statistics and Trends. https://www.invespcro.com/blog/global-online-retail-spending-statistics-and-trends/ (Accessed January 9, 2018).
16. MM&M, 2016. How is Pharma shifting its marketing budgets? Healthcare marketers trend report. http://media.mmm-online.com/documents/214/healthcare_marketers_trend_rep_53381.pdf (Accessed January 30, 2018).
17. Google. 2012 Google's The Digital Journey to Wellness: Hospital Selection. https://www.thinkwithgoogle.com/advertising-channels/search/the-digital-journey-to-wellness-hospital-selection/ (Accessed January 30, 2018).
18. McClure, Dave. September 6, 2007. Startup Metrics for Pirates: AARRR! http://500hats.typepad.com/500blogs/2007/09/startup-metrics.html (Accessed January 30, 2018).

Chapter 5

Marketing Strategy Implementation and Control

Execution of marketing strategy is often more challenging than preparing the strategy itself.

Implementation of your marketing strategy is a series of steps necessary to ensure you achieve your goals. It is a practical approach, the real-world application of your plan. It is also a realistic assessment of how you will deal with roadblocks as they emerge. Meanwhile, control deals with the tools you need to measure your strategies' effectiveness. Without correct execution and implementation, as well as the necessary control mechanisms, your marketing strategy will most likely fail.

Most of the time, this will mean writing a marketing plan. This written document will go over your situation analysis, your proposed strategy, and will detail the activities and actions to be taken to attain your strategic goals. It will include detailed information on who is responsible for specific activities, a timetable, and a list of resources for the activities to be completed.

Finally, remember that planning and execution are two interdependent functions. Hence, successful strategic results are achieved when those responsible for execution are also part of the planning and formulation process.

The next few pages will be dedicated to this transition from strategy to implementation and control.

5.1 Implementation

The implementation of your marketing strategy starts with the planification of the actual steps necessary to operationalize your marketing strategy. The key is making sure that your marketing activities happen in the correct sequence and at the optimal time to ensure success. It is a process, a series of integrated decisions taken over time.

Some estimates suggest as many as 70% of new strategic initiatives fail at the implementation stage. While this number might be skewed a bit high, it is undeniable that incorrect implementation is an issue that needs careful attention.

5.1.1 Implementation versus Strategy

Implementation is a serious shift in your marketing strategy activities. Whereas strategic planning is a controllable process which takes a predefined amount of time (a couple of weeks or months), implementation goes on over a longer period, making it harder to control. Furthermore, as you put your plan into application, uncontrolled events (or the impact of your marketing activities) will create changes in your environment, requiring further adaptation.

Furthermore, going forward with the implementation step will require and implicate more people than were required at the preparation of the strategy, and the more people are involved, the more challenging the execution will be. As such, additional communication across the organization will be needed to ensure that there is a common vision around the marketing strategy, as well as incentives to support the execution of the strategy.

5.1.2 Implementation Plan

Your implementation plan will need to include the following information:

A. The **strategic objective** that is being targeted.
B. The **list of activities** that need to be implemented; these are the activities you will have chosen to focus on to reach your strategic goals.
C. The **responsible** for each of the activity; if everybody is responsible for something, then more often than not, nobody will be responsible for something, as every individual assumes that others will complete the given task.
D. A description of the **resources** allocated for each activity; resources include marketing budgets, specialized personnel, and technical resources. This will often include how each resource will be allocated.
E. A **timetable** which details when each marketing activity will occur; this helps individuals responsible for activities stay on course, and enables coordination between elements which are interconnected.
F. Information on how activities will be implemented; a brief description of **actions** on how each activity will be implemented, to ensure a common understanding by all participants.
G. **Control elements** detailing how the success of each activity will be measured, and how corrections will be made.

A basic framework that includes examples is included in Table 5.1 to enhance our understanding.

Table 5.1 Sample Implementation of a Marketing Strategy

Strategic Objective	Increase Sales by 25% over 3 Months through Promotional Strategy				
Activity	Resources	Action	Timeline	Responsible	Control Elements
Internet advertising	• Budget: US $10,000 to be used on Google AdWords • Marketing coordinator to write ad • Advertising coordinator to post and monitor ad effectiveness	Write and test Internet ad	First week	Marketing coordinator	• Conversion rate (CVR) • Return on investment—digital (ROI-digital) • Cost to acquire a customer (CAC)
		Setup Google AdWords account; post as online	Second week	Advertising coordinator	
		Monitor ad effectiveness;	Ongoing	Advertising coordinator	
Samples program	• Budget: US $50,000 • Product manager to coordinate for sample preparation • Marketing coordinator to coordinate sending samples and monitor responses	Prepare sample size product	First month	Product manager	• Number of inquiries to customer service • Positive response rate ensuing from personalized calls • CAC
		Prepare target list of sample recipients	First month	Marketing coordinator	
		Send samples	Second month	Marketing coordinator	
		Follow-up with non respondents	End of second month	Product manager	

5.1.3 Barriers to Successful Implementation of Marketing Strategy

Knowing the barriers to implementation can enhance the successful execution of your marketing strategies. Barriers fall into two broad categories: external pressures and internal pressures.

5.1.3.1 External Pressures of the Organization

Understanding the external pressures of the organization and how they can impact implementation comes back to the SLEPT (Social, Legal, Economic, Political, and Technological) model.

5.1.3.1.1 Social Factors

Changing patterns in consumer consumption, accelerated aging population, higher education levels, increased health consciousness, and more emphasis on safety are all examples of how social factors can affect your implementation. One of the most significant social factors is the increasing acceptance of social media. Social networking sites and the increased use of mobile phones are both important considerations during implementation.

5.1.3.1.2 Legal Factors

Increased scrutiny of companies in the life sciences field has led to important shifts and modernizations in legal and regulatory frameworks. The way companies communicate with health-care practitioners and consumers can shift rapidly and can impact some of the action steps you had planned in your marketing plan.

5.1.3.1.3 Economic Factors

The overall economic performance can impact consumer purchasing patterns. For example, an important rise in inflation could impact pricing strategy. Furthermore, economic shifts can impact consumers' disposable income, impacting their ability to purchase products.

5.1.3.1.4 Political Factors

In the last few years, we have seen increased government scrutiny and implication in the life sciences industry. As health-care costs increase, there has been increased pressure to reduce the price the hospitals or patients pay for their products. Increased implication in tariffs and duties can also impact companies that rely on international trade for some of their components.

5.1.3.1.5 Technological Factors

Innovation is something that has a deep impact in life sciences. Innovation can impact everything, from the competitive environment to how companies purchase products. It can also reduce barriers to market entry, and can have a very huge impact on manufacturing technologies.

5.1.3.2 Internal Pressures of the Marketing Function

Those relevant in life sciences include leadership, organization model, resource allocation, and lack of incentives.

5.1.3.2.1 Leadership and Management

The degree to which management is implicated in the implementation of the marketing plan will have an important impact on its success. This is especially important if the marketing plan suggests considerable shifts from traditional activities. Opposition may come from individuals without a marketing background and who do not see the necessity of changing the existing marketing activities. Middle management will also often resist changes unless sufficient information was shared or sufficient incentives are offered.

5.1.3.2.2 Organizational Model

We spent some time looking at different models of companies dealing with life sciences. While there are many models, few of these are customer- or market-focused. Hence, for some models, adoption of the marketing strategy will be especially difficult as the organization was never designed to interact with its client base.

5.1.3.2.3 Resources

Your marketing strategy will most likely require the allocation of significant resources (such as financial resources and individuals). As such, successful implementation will depend on these resources being available. The barrier is that either the resources are not available or that leadership considers the other initiatives as more urgent and requires those resources first.

5.1.3.2.4 Lack of Incentives

Rewards can enhance successful implementation. As such, developing specific incentives such as contests, recognition, and extra compensation can be useful to encourage staff to implement and operationalize strategic actions. Performance controls should be fair and encourage participation: setting unrealistic goals will make incentives irrelevant. Finally, focusing incentives on the achievement of the overall objectives of the strategic plans rather than on individual efforts will be beneficial to both team efforts and overall participation.

5.2 Control Elements

Control elements are necessary to ensure that your implementation is advancing successfully. These controls are prepared during the implementation phase, and they give the company something to aim for. They can include measures such as marketing budgets and ratios.

5.2.1 Implementation of Control Processes

Implementation of control processes is usually very straightforward.

Start by setting a value for each indicator you wish to measure. These can be either quantitative or qualitative metrics related with the implementation of the strategy. Quantifiable metrics can have numbers associated with them, such as sales volume, market share, or the number of responses you obtained for your targeted ads. Qualitative factors can include interpretable data, such as answers to a customer satisfaction survey.

You can also determine some tolerance ranges. For example, you could list an objective of 100 widgets sold each month, and that a variation of ±10% would not generate any adjustments in ongoing marketing actions.

Once the marketing activity is ongoing, measure your indicators to ensure you are effectively reaching your goal. Compare your current number to planned values and try to determine the cause of deviations (if any). If you reached your indicators, then you can replicate the activity and expect future success.

If the variations exceed the tolerance levels you had set, you could plan and implement adjustments to reach your initial goals and minimize variations, or you could change the values of the indicators if you found that some changes in the environment has made reaching the goals impossible. For example, a health-care organization I worked with had planned to reach and provide care to 55 families each month. However, loss of key personnel slowed down our ability to match caregivers with families, limiting our uptake to 10 new families per month (instead of the planned 20 new families each month). As such, we had to adjust our control metrics to a lower yet steady adoption rate.

5.2.2 Barriers to the Successful Implementation of Control Procedures

Barriers to successful implementation include inadequate monitoring, inadequate targets, too much management by exceptions, and cost and complexity.

5.2.2.1 Inadequate Monitoring

Control systems that do have correctly assigned personnel or have weak control mechanisms will often have difficulty monitoring results. As such, lack of appropriate monitoring and evaluation procedures will be a significant barrier to the successful implementation of any strategy.

5.2.2.2 Inadequate Targets

The target criteria should be objective and measurable, and will need to be communicated to stakeholders at the beginning of the implementation phase. Targets need to be challenging but achievable, lest they lose all their motivating factors.

5.2.2.3 Management by Exceptions

Some situation will require exceptional resources or will necessitate deviations from agreed upon goals. As such, expectations can be made in cases of *force majeure*, but these should not be the norm. Defining what constitutes an exception to the norm is useful, but even then, these should not be abused, lest it becomes their own norm, generating a set of unusual routines. The best way to attenuate this is to set tolerance levels where deviations are acceptable.

5.2.2.4 Cost and Complexity

A control system can be both costly and complex to set up and maintain. For example, there could be high costs related to implementation such as IT software, hardware, data acquisition, training, and hiring. As such, the benefits of the control systems must outweigh the associated cost, for fear that it becomes an unusable bureaucratic system, becoming more self-serving than customer-focused. Careful deployment and incentives can be the key to ensure the control systems do not become self-centered.

5.2.3 A Word of Caution on Control Systems

It is very important to remember that control systems can stifle effort and creativity. Control systems promote uniformity and conformance to existing targets and can become powerful barriers to innovation. Strict control processes often reduce motivation, creativity, and innovation. Since they imply a perspective of standardization, they are sometimes ill-equipped to evaluate innovation and new improvements. Careful examinations of the causes of deviation become necessary when innovation is being deployed on day-to-day processes, and new measurement methods and benchmarks might become necessary.

An example would be a process which would improve online inventory accuracy, which, in the long term, would improve online sales, but which reduces individual sales numbers (as the staff spend a bit more time updating inventory, rather than focusing on increasing their own sales number). The whole system would benefit from increased awareness of the importance of an accurate inventory but could be unable to integrate the benefit in existing metrics.

Also, control systems promote inspection rather than development. As such, they will often have problems diagnosing systemic issues with processes, often focusing on short-term fixes rather than identifying systemic issues. This leads to a philosophy of constant firefighting (focusing on quick fixes to specific problems), ignoring longer-term issues, and limiting initiatives which would radically improve a dysfunctional system.

Chapter 6
Marketing Metrics

Metrics are quantitative measures that an organization can use to measure, compare, and track performance. In marketing, they are used to measure a series of activities such as sales growth, revenues, and advertising efficiency. They are often compared to industry benchmarks to ensure that a company is performing adequately, and that it is reaching its strategic targets. Measuring the impact of different marketing activities is essential to ensure optimal resource allocation in an organization, as it enables decision-makers to understand the relationship between strategic variables and executional variables.

In this section, we will go over some of the basic ratios life sciences companies can use once they reach commercialization. The ratios we will be looking at range from the more classical (such as return on sales) to the more complex ones (such as advertising-to-sales and customer acquisition cost [CAC]), as well as few specifically for pre-revenue organizations. We will close the chapter with a few ratios specific to digital marketing campaigns which measure traffic, conversion, and revenue.

6.1 Why Use Metrics?

The purpose of establishing marketing metrics is to operationalize the firm's strategies and to track the impact of those strategies on target customers and the company's financial objectives. Using metrics allows a company to quantify the impact of its strategy, and in doing so, it can better understand the relationships between strategic variables and its execution. As such, metrics are useful to diagnose issues and improve performance in a continuous manner, diagnose failures in marketing executions, and enable the organization to prioritize initiatives and issues that need the most attention.

6.2 Some Pre-revenue Ratios

The pre-revenue ratios we will be reviewing can be used before commercialization, for planning purposes. As such, we will be reviewing sales force coverage and break-even analysis.

6.2.1 Sales Force Coverage

As you reach the commercialization stage, you will need to hire specialized sales staff. Even at the planning stage, it might be useful to start to forecast how many individuals you will need to hire, if this is the route you choose in your commercialization effort. Each of these individuals will be responsible for a territory (based on a geographic region, sales potential, history, or a combination of these factors).

The purpose of this formula is to create balanced sales territories and align expectation internally. If a territory is underserved, it might lead to clients being underserved as well, and lead sales staff to engage in negative behaviors (identify fewer newer prospects, spend too little time with current clients). To balance this, you can use the following formula:

Workload (hours) = [Current accounts (#) * average times to service an active account (#)] + [prospects (#) * time spent trying to convert a prospect into an active account (#)]

where

- Workload is the number of hours the individual is expected to work on client-related activity.
- Current accounts is the number of client accounts the individual is usually responsible for.
- Servicing an active account includes issues such as handling complaints, account maintenance, service meetings, as well as responding to special requests and collection;
- Prospects is the number of leads the salesperson is expected to maintain.
- Time spent trying to convert a prospect into an active account is the time spent each week working through the sales funnel.

So, in a situation where a company estimates that each week existing clients generate, on average, 5 hours of work, and a prospect takes about 6 hours to develop, you could optimize your salesforces accordingly:

35 hours = 5 hours + (x * 6 hours) = 5 new prospects

Note that the sales representative's time is not 100% covered by that workload and that some time should be allocated for items such as planning, administration, and order processing. Usually, you would allocate 10%–15% of a persons time to these activities.

6.2.2 Break-Even Analysis

The break-even analysis is a ratio that falls on the cusp of marketing and financial responsibilities. The objective is to calculate at which levels the sales generated by the company cover all the costs (both fixed and variable), so that you are neither losing nor making money. The formula is as follows:

Sales = Fixed costs + (variable costs × number of units sold)

Break-even analysis is built around three key assumptions:

a. **Average price per unit**: Determine the average price at which you are selling a unit. Include exceptional events (sales and discounts) into your average price if necessary. If you have more than one product for sale, you might need to average these out, or use scenario-based estimates.
b. **Average cost per unit**: This is the cost to produce each unit. If you are offering a service, it is the cost to provide the service on a per activity basis.
c. **Fixed costs**: These are the costs to operate the business, even if no products or services are sold. These include salaries, rent, equipment, and administrative fees.

If you have a problem calculating price and costs per unit, and are running a fairly standard set up, you can use a percentage estimate, running at a 50% margin (average cost per unit 0.5, average revenue per unit of 1). It will give you a rough estimate of your break-even point.

As an example, imagine a small medical device company that has a fixed operating cost of $250,000 per year, a variable cost of $2.5 per widget (cost to produce and sell each widget), and which sells its widget at $5 each. You would get the graph as shown in Figure 6.1.

The break-even point is exactly at 100,000 units. Any fewer units sold means the company is operating at a loss, and each unit over 100,000 units starts generating a profit for the

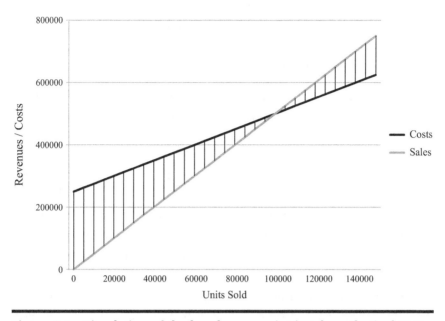

Figure 6.1   Simulation of the break-even point in a hypothetical SME.

company. Hence, the start-up founders can ask themselves the following: Can I realistically sell 100,000 units each year? If not, can I make changes in my costs (either variable or fixed) so as to reach the break-even point earlier?

Break-even is not as commonly used nowadays, since it relies on sometimes crude estimates for your costs per unit and your fixed costs. Furthermore, it is sometimes confused with payback period, which it is not. You do not include the investments to start your company, or the R&D to design your product. You focus exclusively on the costs to operate your organization, and the costs to manufacture and sell your products/services.

Nonetheless, break-even analysis is a useful tool to estimate minimum sales required to operate. If you are starting out, it might be more useful to calculate using more conservative numbers, to generate a conservative estimate. You could always calculate it again once you are in operations, with more precise numbers later on.

6.3 Ratios to Measure Sales Effectiveness

6.3.1 Return on Sales

Return on sales (ROS) is a financial ratio that is used to monitor how efficient a company is at generating profits from its revenue, analyzing what percentage of total company revenues is actually converted into company profits.

Decision makers are interested in the ROS because it shows the percentage of money that the company actually makes on its revenues during a period. They can use this calculation to compare company performance from one period to the next.

The ROS formula is calculated by dividing the operating profit by the net sales for the selected period. For example, a company that generates $100,000 in net sales in a month has $90,000 in operating costs each month; hence, generating $10,000 in operating profits would have an overall ROS of 0.1. As such, the more efficient the company is, the higher the ratio.

6.3.2 Advertising-to-Sales Ratio

The advertising-to-sales ratio is used to measure how effective advertising was in generating short-term sales. This is calculated by dividing the total advertising expenses by sales revenue and measuring the impact over time.

An advertising-to-sales ratio that is diminishing over time would demonstrate that advertising expenses are generating new sales; this could mean the campaign was successful. A stable or increasing ratio indicates that the advertising campaign did not successfully generate sales. To better understand, let us go through an example.

A small company has decided to start an online advertising campaign, spending $5,000 each month (Table 6.1).

In our example, we can verify and confirm the effectiveness of the advertising campaign.

Table 6.1 Application of the Advertising-to-Sales Ratio

Period	Q1	Q2	Q3	Q4
Monthly advertising spending	$5,000	$5,000	$5,000	$5,000
Monthly sales	$100,000	$110,000	$120,000	$125,000
Advertising-to-sales ratio	0.050	0.045	0.042	0.040

Note that some advertising campaigns are designed to generate long-term sales, or additional visibility, so looking at this ratio through short-term lens might not reflect the long-term benefits of the advertising.

6.3.3 Customer Acquisition Cost

This ratio is used to calculate how much you are spending to acquire each new client. To calculate it, add up your total sales and marketing cost and divide it by the number of new clients acquired. So, add up all the spending related to advertising, the relevant salaries, commissions, and bonuses, as well as add an overhead within a given time period, and then divide it by the number of new customers that you acquired in that same time period.

For instance, if you spent $15,000 on sales and marketing in a month and added 20 new customers that month, then your CAC is $750. You can then monitor this cost per customer over time. If it rises, it could be indicative that the strategies you are employing are losing their effectiveness and that some adjustments are warranted.

6.3.4 Marketing Percentage of CAC

To calculate the marketing percentage of CAC (M%-CAC), you simply calculate your marketing costs, and divide that number by the total sales and marketing cost, multiplied by 100. So, using our previous example, if the company calculated a marketing cost of $3,000, divided by the total sales and marketing cost of $15,000, it would have an M%-CAC of 20%.

By calculating the share of marketing in the CAC, you can keep tabs if you are spending too much on marketing or too much on sales. If you depend mostly on an outside salesforce that deploys a long and complicated sales cycle, M%-CAC might be only 10%–20%. If you have an inside sales team and a simple sales process, M%-CAC might be closer to 20%–50%. And if you have a low sales cost and very simple sales cycle where sales are almost automated, the M%-CAC might be closer to 60%–90%.

6.3.5 Average Retention Cost

This ratio is used to calculate how much is being spent on customers to retain them. While it can be challenging to isolate direct costs related to retention, monitoring these costs on a per customer basis can help identify clients that are costing more to the organization than they are generating.

The average retention cost is usually calculated by isolating retention costs (from CACs) and dividing by the number of clients retained. As such, the firm needs to monitor the number of customers that it is attempting to retain and the total number it did manage to retain.

An additional source of difficulty is that some customers will be retained, even if no money is spent on them. As such, this ratio can be a little misleading, and increasing it will not necessarily increase customer retention.

6.3.6 Lifetime Customer Value

Calculating the value of each customer can be useful if you are trying to forecast revenue and growth, as well as comparing these numbers to customer acquisition and retention costs. You can calculate the lifetime value of your customers by multiplying the average sale per customer by the average number of times a customer buys per year by the average retention time in months or years for a typical customer.

Some ways to increase the lifetime value of your customers include preparing campaigns that reach out to existing customers and updating existing customers about new services and products (Table 6.2).

Table 6.2   Marketing Metrics Cheat Sheet (If You Are Generating Sales)

Metric Name	How to Calculate	What Does the Ratio Mean
Return on sales (ROS)	Divide the operating profit by the net sales for the period	If the number increases, the company is becoming more efficient at generating revenues
Advertising-to-sales ratio	Divide total advertising expenses by sales revenue	A high advertising-to-sales ratio indicates that high advertising expenses resulted in low sales revenue. A low ratio may indicate that the advertising campaign generated sales.
CAC	Add up your total sales and marketing cost, divide it by the number of new customers	If it increases, it means that you are spending more to acquire new clients; this should be monitored over time
Average retention cost	Identify specific retention costs, divided by the number of clients retained over a set period of time	If it increases, it means that you are spending more to retain your clients; this should be monitored over time

(*Continued*)

Table 6.2 (*Continued*)   Marketing Metrics Cheat Sheet (If You Are Generating Sales)

Metric Name	How to Calculate	What Does the Ratio Mean
M%-CAC	Add all marketing costs, divided by total sales and marketing costs	You should target: 10%–20% if you have a long and complicated sales cycle; 20%–50% if you have an inside sales team and simple sales process; 60%–90% if your sales process is automated
LCV	Multiply the average sale per customer with the average purchase per year per customer and the average retention time	If the ratio increases, your customer value is increasing and you are reducing customer churn; this should be monitored over time

6.4 Digital Marketing Metrics

Your digital marketing efforts and advertising will require distinct metrics to measure. Fundamentally, there are three types of metrics you can measure: traffic metrics, conversion metrics, and revenue metrics.

6.4.1 Traffic Metrics

Traffic metrics are used to measure and monitor the number of users visiting your assets during the traffic generation stage of your campaign. At this stage, you should be focusing on visibility and overall presence.

6.4.1.1 Overall Site Traffic

Your website's overall traffic is one of the easiest metrics to measure and monitor. Shifts (either upward or downward) can give you some basic insight on whether your overall campaign was a success. Some methods of measuring site traffic include the following:

- **Page views**: The number of times a page was loaded. This indicates how many times somebody reads your website. You can also see which pages garner the most views and adjust your content accordingly.
- **Unique visitors**: The number of unique visitors that visited your web page. Of note, this accounts only for the number of different "computers" that visited your website, not individuals. The accuracy of this metrics is debatable, but it is a good general indicator of how many different people are traveling to your web page.

6.4.1.2 Monitoring the Source of Web Traffic

Monitoring the source of your traffic can give you valuable clues as to why users are coming to your website and what keywords they used to find you. There are different elements you can monitor:

- **Source of the web traffic**: While most of the traffic you drive will come from search engines, it is possible to generate traffic from a variety of sources such as paid search, social media posts, social media ads, email marketing, content and blogs, and links from other sites. Identifying which sources are generating traffic can be useful to see if elements of your marketing strategy are successful.
- **Type of web traffic**: As users increasingly use mobile devices to surf the web, the growth in mobile traffic

continually increases. As such, it might be useful to monitor if your traffic is being generated through traditional devices or if they are visiting your digital assets through mobile devices (such as their smartphones and other Internet-capable mobile devices). This metrics can offer insight on how you can successfully structure your content, ensuring better engagement with both mobile and non-mobile website visitors, as well as optimizing how your digital assets are displayed online.

6.4.1.3 Monitoring the Paid Traffic

If you paid to enhance your traffic (through sources such as Google Ads), then you will need additional and specialized ratios to analyze your traffic, to ensure that the traffic is generated by these paid advertisements, and is not being generated organically.

- **Cost per click (CPC)**: CPC defines the amount of money you are paying to get a click on your ad. This will vary for the popularity of the word you choose, so choosing something generic like "lab services" will be a lot more expensive than "molecular pathology."
- **Click through rate (CTR)**: CTR measures how many people viewed your advertisement online, and then clicked on in to visit your website. Hence, it is an efficacy ratio, where you can see that the higher the CTR, the more people were the interested by your ads. This can also result in obtaining rebates from search engine marketing platforms such as Google AdWords due to the perceived high quality of your ad.

6.4.2 Conversion Metrics

The second set of metrics deals with the conversion of viewers, from interested to potential leads. As discussed earlier, converting website traffic into business leads or sales is the primary purpose for your digital marketing campaign. As such, metrics that provide insight into a successful campaign are important during conversion.

- **Conversion rate (CVR)**: CVR is one of the most important metrics used to measure the success of your digital marketing effort. It is the percentage of unique visitors that complete a desired action on your website. Conversion goals can include signing up to receive a newsletter, contacting customer service to set up an appointment, or even an actual sale. As such, measuring how many visitors convert into leads is an important function, and can indicate opportunities to optimize your digital marketing. For example, you could create two distinct market campaigns and compare the CVR on them to choose which one to continue (often called split testing). In health and medical, CVR is usually around 2.5%. Lower conversion rates could be related to poor website design or disinterested visitors.
- **Bounce rate (BR)**: The BR displays the percentage of visitors who leave your website after visiting a single page, without clicking any other links. In most cases, they will have found your content irrelevant to them. To lower your BR, you should start by looking at the traffic origin of your bounces, to see if some marketing channels are inappropriate and need optimization. If the problem is widespread, your overall content might need optimization.
- **Average page views per visit/average time on site**: These two indicators indicate how much time and content visitors consumed on average. Short visits could indicate lack of interest, whereas long visits could be indicative of attentiveness. Furthermore, the longer the visits, the more chances for engagement and conversion in leads/paying customers.

6.4.3 Revenue Metrics

The final goal of any digital marketing campaign is to generate sales. To quantifiably examine your web success, you should measure it using revenue metrics. Doing comparable testing (multiple campaigns) will enable you to find which campaigns are successful and which need optimization.

- **Return on investment—digital (ROI-Digital)**: The digital ROI is measured by calculating traffic that converted into paying customer compared to the overall traffic generated. This will indicate the overall return of your campaign, relative to the amount directly invested into it. This is obtained by calculating revenues generated by new clients online during a certain period divided by the overall cost to acquire them.
- **Cost to acquire a customer (CAC)**: While ROI will measure the number of new customers, it will not indicate the total cost to acquire each of these new customers. CAC calculations (total cost of digital marketing campaign divided by the number of new clients during the same period) will calculate the cost to acquire this new client and give you some insights into the profitability of your endeavor, especially when compared to the LCV of your customer. Depending on your marketing strategy and goals, if your CAC is higher than your LCV, some optimization is definitely warranted.

6.5 Final Notes

A few closing words on metrics before moving on to our last chapters.

For one thing, the usefulness of metrics is based on the quality of the data used. Using imprecise or incomplete data will hinder, not enhance, decision-making. As such, decision makers have to be especially cautious of the data they collect and use for their decision-making.

To conclude, it is important to understand that health-care is especially vulnerable to making decisions using ratios and metrics. The more you deal with human beings and the less you deal with objects, the less useful metrics should become. Over-rationalization of irrational elements seldom delivers great results for those involved.

Chapter 7

Discussion on Unique Perspectives

The majority of the previous sections dealt with building a marketing strategy for a traditional life sciences product—a physical device or a typical drug, for example. Although we did address some unique cases through some specific examples in the book, there are three specific perspectives that we have not sufficiently addressed and that we will be doing so right now.

First, we have included more information to emphasize the specific challenges relative to life sciences marketing, relative to marketing in general. This is followed by a short section on the specific challenges related to the marketing of health-care services, and how this is very different from marketing health-care products. Finally, we will be going over the specific challenges related to formulating a marketing strategy for digital products and services.

7.1 Marketing in Life Sciences

Marketing in life sciences is not as simple as marketing normal consumer goods. The products and services are very complex;

they have a direct impact on health, and they address a population that does not always have the knowledge to properly evaluate and distinguish claims. Hence, your message and how you deliver it must balance the information and educational content without being patronizing and insulting. It is a subtle balance to achieve.

Furthermore, companies in life sciences have to contend with regulated advertising frameworks, and they must conform to advertising guidelines. While in this book we cannot go into the specific laws and regulations that apply to every country, some rules of thumb include the following:

- Do not present anything in your marketing materials that is not true or attempts to mislead and/or misinform the public.
- Do not directly disparage a professional colleague/competitor.
- Do not state anything in your marketing that makes you appear better than your competitor.

Some of industry associations/documentation you may wish to consult are given in Table 7.1.

Also, you have to take into consideration that there is a knowledge gap between you and the consumer, especially if you are targeting everyday consumers (rather than targeting health-care personnel, for example). Some of the terms and notions that you will usually utilize internally, while very fundamental for you, may be complex and unfamiliar for them. As such, remember to shift from a more traditional health-care language to one the consumer can relate to.

Another good practice to remember is that companies in life sciences often have an overall objective of helping patients live longer, healthier lives. Therefore, try to focus on this in your marketing strategy as part of the message, focusing on how you can partner with patients to help them enhance their health.

Furthermore, it might make sense to build a marketing strategy that targets the population around the patients (such

Table 7.1 Resources for Advertising Guidelines Directly to Consumers (B2C)

Industry Association	Relevance
Pharmaceutical Research and Manufacturers of America	Publishes direct-to-consumer advertising principles for its members
www.phrma.org	
Pharmaceutical Advertising Advisory Board	Offers preclearance review services that support trustworthy health product communications that comply with the Canadian regulatory framework
https://secure1.paab.ca	
European Observatory on Health Systems and Policies	Pharmaceutical advertising regulatory framework applicable in 15 European countries
http://apps.who.int/medicinedocs/documents/s23163en/s23163en.pdf	
Japan Pharmaceutical Manufacturers Association	Code of ethical practice in advertising for pharmaceutical companies
www.jpma.or.jp/english/parj/pdf/2017_ch02.pdf	

as caregivers and the family) since they might be more receptive to your message, its advantages, and its technology. For example, imagine that you are selling a tracer bracelet for patients that allows you to follow in real time where the patient is. The bracelet alerts the caregiver when the person leaves his house and shows were the patient is going. Now, in your marketing plan, the target user could be people with Alzheimer's disease, but it could be challenging to convince them of the benefit of your device. But by targeting the families around those patients, you can stress the benefits of your device quite handily; hence, it might make a lot more sense to target them through a digital advertising campaign, even though end users (people that have Alzheimer's disease, which are usually in older customer segments) are not traditional Internet users.

Overall, you should focus on the positive elements of services and products that you offer, being honest about what you provide and what your experience is.

7.2 Marketing Health-Care Services

7.2.1 Difference between Services and Products

Before discussing the specific service strategies a company can deploy when selling health-care services versus products, we should take a moment to document the difference between services and products. The traditional way to distinguish the two is to focus on the aspect of physicality. Services are usually intangible in nature. On the other hand, products are tangible and discernable objects that you can grab, touch, and feel. As such, this might be the best way to distinguish them. Products are objects that you can interact with using your five senses, while you cannot do the same for services. For a consumer, purchasing a service is somewhat different than purchasing a product; it is hard to try out a service before purchasing it, whereas a product can be held, touched, and observed prior to a transaction.

However, this description is quite inadequate when dealing with digital objects (such as applications and software). The debate on where digital objects land on the service/product spectrum has been ongoing for years, and the deployment of Software as a Service has only increased the debate. Rather than attempting to resolve it in this volume, I have added a distinct section (Section 7.3) which deals with digital products specifically.

Also, as companies engage in innovative marketing strategies, the line between services and product increasingly blurs. For example, many companies now offer service–product combos (e.g., a device, with a 1-year subscription to the accompanying software), making the distinction more complex than just black and white. In fact, today, most products come with a service component (warranty, service contract, etc.), and strategies must be built to reflect this duality.

Health-care services include all types of services where care is provided to the patient. This can include primary care, hospital care, and dental care all the way to long-term care. It can also include a whole host of services such as clinical nutrition, music therapy, occupational therapy, psychology, and physiotherapy. It can even include services such as lab analysis (as the service being offered is the analysis, not the equipment). In developed countries, health-care services occupy a large part of total spending, accounting close to 50% of all consumer health spending.

7.2.2 Developing Marketing Strategies for Health-Care Services

On the surface, companies selling services employ most of the same marketing strategies as manufactured goods. Per se, these companies engage in similar strategies such as differentiation, they compete on price, or they create branding strategies. As such, the basic assumption is that the marketing services will follow a lot of the same strategies that physical goods do. Health-care service providers must build their marketing strategy on the same four pillars of marketing: product, price, distribution, and promotion. Furthermore, the use of market research remains an essential component of building a marketing strategy. Nonetheless, some difference will be observable on the application of different strategies. We will go over these when tackling the topics of intangibility, fluctuating demand and perishability, client–patient relationships, and uniformity of service.

One of the biggest differences between services and products, the notion of **intangibility**, was discussed earlier. Products are usually tangible, whereas services are most often intangible. Being intangible, it is difficult for marketer to appeal to the consumer's five senses in promotional materials. This creates new challenges for them. Fundamentally, there are two main strategies to get around the intangible nature of services:

1. First, putting the emphasis on the physical elements that are present in the service offering. That way, it will be easier for the consumer to go from an abstract concept to a concrete artifact. For example, a company selling superior quality dental services could focus on the modern technology equipment, the superior components used, and the professional staff to grab the patients' imagination. The client can then convert the intangible into something more tangible, and be more comfortable in the transaction.
2. Second, the end results of the intangible service should be emphasized in the marketing materials. If a company offers physiotherapy services that offer rehabilitation services for high performing athletes, and they have a high level of success with customers, they should emphasize these successes in its promotional materials through case studies, videos, and success stories.

Another important difference between products and services is the **fluctuating demand and perishability**. Unlike physical goods which can be stored and consumed later, when a service is not used at a specific time, it "perishes." For example, imagine that you have set up a convenient mental health counseling service where people can call in to discuss their mental health issues and obtain assistance. You will soon find that you have fluctuating demand (most people call on weekends or after work), and there are long periods of time when the counselors are not very solicited (e.g., early morning or late at night). When the counselor is not answering the phone, he is not using his time to provide services, he cannot "store his time," yet he is still paid. Hence, the resource (the counselor's time) is perishing. In the same way, a consumer cannot call a line and "save" an hour of conversation for use later. Once he calls in, he must use the hour right away.

In situations where your product is especially susceptible to fluctuating demand and perishability, there are two main strategies you can employ. First, you can emphasize **off-peak pricing**. This strategy is used to create different price categories for your services, making sure prices are less expensive during off-peak hours and more expensive during peak hours. Usually, this will translate into a more distributed usage curve—some people will prefer calling off-peak in an effort to reduce cost, whereas others will not mind the extra cost and will call whenever it is convenient for them.

The second strategy is to invest in automated technologies that allow you to optimize service delivery; this could include the following:

1. Using artificial intelligent agents to handle calls as a first line of contact (which could be simple requests), and referring complex requests to a reduced number of counselors. An example of this is Babylon Health, a UK-based start-up that has developed a technology which is a mix of artificial intelligence and video/text consultations with doctors. In 2017, it did a 6-month trial to offer AI-powered chatbot "triage" as an alternative to telephone helpline to answer simple requests and refer patient to pertinent resources [1].
2. Use of automated voice mails offers the option to take a message, explaining that an agent will call them as soon as they are available. If the person insists, the person would be put on hold for the next available agent. Alternatively, you could even design the system to let the client set the time that they want to be called back.

In both the cases, the idea is to distribute the workload over time, reducing peak waiting times and the product's (client or service agent) time that is perishing due to nonuse.

Another important factor in the service sector, especially health-care services, is the **client–patient relationship**. The person providing the service is seen by the client as a trusted expert. Indeed, in many cultures, doctors and nurses are perceived as friendly experts and as such, there is a powerful relationship of trust which is built between the patient and his health-care provider. A breach of trust can create difficult challenges for health providers, so it is best to emphasize the expertise of the health-care provider.

One of the ways to emphasize this is to focus on storytelling. Write with empathy and focus on accuracy. By placing the health-care provider at the center of the patient story, sharing information and educational materials can go a long way in promoting your facility in the eyes of patients. Furthermore, by focusing on the information that patients need, you are more likely to appear on their research results, further contributing to their positive image of you. You might also profit from social media component that facilitates communication with health-care personal, being able to answer simple questions.

One last important issue is **uniformity of service**. When producing manufactured goods, it is quite simple to make sure that the goods are all the same, and that they all respect the same standards. As services are provided by individuals, it is quite challenging to ensure service uniformity. If customers get uneven services at different times, they will judge your service by the lowest service quality they receive, not the best one.

While carefully selecting the caregiving personnel and accompanying personnel integration with extensive training can help your organization, your quality efforts must be sustained by internal marketing efforts, where staff is continually encouraged, stimulated, and supported in their work. Pitching in to help out when there is overflow, words of encouragement and properly explaining to employees the importance of their work can play an important role in ensuring a constant level of service. Finally, engaging in mystery shops (where you employ third parties to replicate the customer experience through the eyes of the customer by simulating a customer purchasing process) can play an important role in identifying opportunities for improvement.

The results of a mystery shop can be very revealing. A client of mine had asked me to contact multiple locations selling his products, and compare factors such as speed of response, quality of response, and follow-up following a first quotation. He was astonished when we went over the results, and learned that some of the locations I had contacted had either taken three times to respond to my request for product information, or even worse, had never responded to inquiries. He was able to identify weak distribution locations and remove them from his distribution channels, and work with functional locations to ensure they continued to offer high-quality responses to potential clients.

Once you are able to provide a constant level of good service, this should reflect through positive word-of-mouth as well as positive feedback in the different digital portals/forums that exist.

7.3 Marketing Health-Care Digital Products

Digital health-care objects are on the cusp of the definitions of both services and products. While digital "objects" are intangible, clients are able to use them at the time of their choosing. Also, many digital products enhance the doctor–patient relationship. As such, some companies attempt to recreate the dichotomy by making the distinction between digital goods (which are still intangible products) and digital services (which are offered online).

Interestingly, digital goods are often perceived by consumers as having lower value that their physical counterpart. For example, digital books will often sell for less than its physical counterparts. While some argue that this is due to the digital product costing less to manufacture, I believe it has more to do with the perceived ownership of the digital goods. You own a physical good, and its ownership is not subject to the existence of a third party. Furthermore, you can resell your physical good without complication or issue. Overall, the perceived value of digital goods is diminished by its intangibility, as well as the fact that its existence is related to either a third party (the company that designed and is hosting it) or the device the product was purchased on (if your cell phone dies, it can be a complex operation to migrate digital goods you purchased to your new device, especially if the operating software no longer supports it, or if the company has stopped selling/distributing it).

While marketing services and digital space have a lot in common with traditional goods, they each have their own sets of challenges. Many problems can be traced back to the intangibility of goods or to the uniqueness of the delivery. As such, it is important that service providers and digital product providers be able to appreciate the commonalities and differences.

7.3.1 Developing Marketing Strategies for Digital Products

In digital health-care, it is very important to identify **multiple revenue strategies** and find different verticals that can be targeted. Diversifying risk is much easier in this space as some of the regulatory hurdles do not always apply, and they do not have extra costs associated with it when developing simultaneous opportunities.

Digital products are **easier to customize** than their physical counterparts, so it is conceivable to create many different versions, each one adapted to a specific customer segment. While new physical product segmentation will have to justify a certain segment size to explain the commercialization of a specific product, digital products have lower upfront costs to enable customization. As such, companies can design specific user versions for multiple different publics. For example, in a traditional physical product, you will often have a standard product sizes (small, medium, large), while a digital product is able to customize different versions of its digital good to many different publics (it is not uncommon for digital products to have four versions or more of the same product, each one targeting a distinct market). In one case, a client of mine had planned six product categories, including members, healthcare personnel, students, elderly people, government officials, and international users.

The second distinction is the ease of adjusting the product to current market conditions, through either **unbundling features** or the **creation of product bundles**. With physical products, creating a new product bundle can be an expensive endeavor and leads to a slew of distinct physical challenges (new packaging, distinct storage, distribution issues). In digital applications, it is possible to bundle and unbundle products following a more flexible approach (e.g., client request for a specific bundle). You might even be able to partner with another product creator, bundling two complementary products to provide a superior customer experience. These collaborations could range from supplementary partnerships (providing your product as the lead product, with access to a secondary product as a feature) to full partnership (both products are sold together by both providers as a unique new bundle). Also, it is possible to convert or create new derivative products with more (or less) features, and have the product available in short order and for a limited timeline according to market requirements.

Another successful strategy for digital goods and services is the instauration of a **trial period**. As the digital goods can be "turned off" remotely at a pre determined date, without any physical actions from either the seller or the purchaser, it is quite easy to let the user have access to a full version of your digital good for a limited period, so they can test the product fully and inquire if they have questions about it. Some companies successfully pair a well-executed communication campaign with this strategy, sending daily/weekly reminders to users during the trial period, sharing tips and tricks, or offering seminars on how to effectively use the digital product.

The length of the trial is up to you, but it should be long enough for the potential customer to be able to evaluate your product without feeling rushed. If your product is relatively simple, a shorter trial (5 or 14 days) might make more sense. If your product is more complex, a longer (30 days) period might be the one to choose. If the customer is still unsure after the trial period, it might even make sense to extend it once, to increase chances of a sales conversion.

Finally, the purchase of digital goods can often be paired with additional bonuses that raise its perceived value for customers. For example, it is quite easy to include an ancillary product (an electronic book on the topic of interest) with each purchase, or give access to exclusive discussion groups where other clients discuss the product, share tips, and get support when facing installation or issues. It is also possible to include extra content, access to beta versions of newer digital goods, or offer one-on-one coaching/30-minute consultation, following the first purchase.

7.3.2 Issues with Health-Care Digital Products

At the time of this writing, the deployment of digital products in health-care settings remains a very challenging topic. In many locations, there are no regulatory pathways for these products to gain official authorization to reach the market. Even in the more innovative environments, the clinical trial pathway that these products should use to be "validated" does not exist. Yet, the functions they perform are getting increasingly complex, from being simple reference tools to performing simple diagnostics and first-line care.

As such, the whole discussion around **liability** is one that remains to be tackled. For example, in some countries, it is possible to download a digital application directly to your phone that will "diagnose" a skin condition, identifying the presence of skin cancer. But if the digital application does not detect a skin cancer in a patient, who later dies from it, who is responsible for the death? Is it a "buyers beware" situation? Reversely, if the application "plays it safe" and overreports potential cases, leading to an increase in doctor appointments, what use is the device? This increased risk, for patients, doctors, and companies in general, remains an issue of contention.

Security of the data collected is another important concern. Many new digital products send the data they collect back to the service provider. This health data is both very valuable and private. As such, being hacked by a third party could lead to extortion, or even lawsuits for breach of policy, and this means that the security of your product is a careful consideration when both designing and commercializing products at large.

Also, many health-care settings are **conservative**, and are slow to adapt new technologies, especially if they cannot obtain official certification. As such, digital products face even more resistance in some settings, as these new technologies pose potential safety problems if the users and caregivers are not properly trained. In these cases, emphasis on successful product integrations and pilot programs, superior training solutions, and identification of a local technology champion is necessary as part of a marketing strategy.

Finally, as products in digital goods often address customer needs in new and interesting vectors, it will be most likely that your potential customers will not be aware of your product. They simply have never used a digital product to tackle their problem. As such, **product discovery** is an increasingly big challenge for companies developing digital products in health-care, and your marketing strategy will need to include components that communicate, share, and make your product known quite well. Emphasis on product recognition is the key for digital health-care products.

7.4 Final Notes

Unique products and services warrant their own chapter, so the information provided here is more of a thought starter than definitive answers. The objective of this chapter was to spark conversation and encourage you to search for more information. As IT and health-care companies increasingly integrate and codevelop technologies, we can expect these different issues to increasingly come into play.

Reference

1. O'Hear, Steve. Babylon Health Partners with UK's NHS to Replace Telephone Helpline with AI-Powered Chatbot, January 4, 2017, *TechCrunch*, https://techcrunch.com/2017/01/04/babylon-health-partners-with-uks-nhs-to-replace-telephone-helpline-with-ai-powered-chatbot/ (Accessed December 27, 2017).

Chapter 8
Final Thoughts

One of the important trends to keep an eye on for in life sciences over the next few years is the role of patients in disrupting traditional health-care systems.

A conventional health-care relationship is usually defined as one between a patient and his doctor. But today's complex health-care environment integrates several stakeholders, each with their own interest. And sometimes, those interests are disengaged from the patient's interest. Add to that increasing costs making health care unaffordable for some patients and increasing congestion in accessing health care, and you have a recipe for disruptive innovation.

As such, it is not surprising to see the commoditization of health care and the growing trend of patient self-care.

For many patients, the Internet is rapidly becoming the primary source of information to go to when there is a health issue, and while the quality of information can greatly vary, the patient often does not have the understanding to evaluate them. As such, we have seen the emergence of the "patient–doctor" meeting a health-care professional (doctor, nurse, pharmacist), armed with printouts of possible diagnostics and potential solutions, challenging the professional's expertise based on the information of sometimes questionable value.

Simultaneously, this has led to the rapid growth of the self-health sector, encouraging the emergence of technologies that facilitate patient self-diagnostics. If a few years ago, these technologies were limited to taking your pulse and measuring your pressure, today's apps can turn your mobile phone into a skin cancer diagnostic device, or enable you to monitor your diabetes, asthma, depression, celiac disease, blood pressure, chronic migraine, pain management, or menstrual cycle irregularity. The growth of diagnostic devices and apps that attempt to reproduce the doctor's diagnostic role demonstrates the interest that patients have in short-circuiting the health-care system and taking back their health into their own hands.

The Theranos story, where blood tests were commoditized and were sold in retail locations, demonstrates how far patients are willing to go to disrupt the traditional health-care system, redefining their relationships to health care.

Add to that trends such as medical tourism (where patients, looking to save costs, think nothing of adventuring half-way across the world in search of care, when the distance they would travel was limited to 40–50 miles at the beginning of the century), "doctors on your phone" (apps that put you in contact with a physician simply by clicking on the phone for a web-based consult), and the rise of nutraceuticals (posting 7%–8% CAGR growth rates each year), and you have the recipe for a potential disruption of the traditional healthcare space.

Some key opinion leaders don't believe that the disruption will be as profound. Yes, new technologies will allow the patient to increasingly take care of himself, but he won't be using the tools to increase self-care, rather he will use them to enhance his relationship with his doctor. The additional information he will be generating will help physicians understand the patient better, and help them in providing care. Even then, the challenge might come in doctor's acceptance of all this user-generated data, and having buy-in from the top might be the best way to ensure adoption.

As technologies disrupt the life sciences space, companies will need to constantly refine their marketing strategy, leading to either newer business models that focus on improving doctor–patient relationships, or models that enhance a customer's experience, encouraging them to reintegrate new upgraded health-care models.

Overall, it is my hope that this book assists you in your journey into this fast-changing, disruptive market.

Bibliography and Further Reading

Aaker, David. 2013. *Strategic Market Management*, 10th edition, John Wiley & Sons, Hoboken, NJ.

Anderson, Eric, & Simester, Duncan. 2003. Effects of $9 Price Endings on Retail Sales: Evidence from Field Experiments. *Quantitative Marketing and Economics* 1: 93. https://doi.org/10.1023/A:1023581927405 (Accessed January 30, 2018).

Ansoff, Harry Igor. 1970. *Corporate Strategy*, Penguin Books, Great Britain.

Bhanji, S. Price. 2012. Discrimination in Pharmaceutical Companies: The Method to the "Madness." *Harvard College Global Health Review*, www.hcs.harvard.edu/hghr/online/price-discrimination-method/ (Accessed January 29, 2018).

Bizer, George Y., & Schindler, Robert M. 2005. Direct Evidence of Ending-Digit Drop-Off in Price Information Processing. *Psychology and Marketing* 22(10): 771–783. doi:10.1002/mar.20084. https://doi.org/10.1002%2Fmar.20084 (Accessed January 9, 2018).

Court, David, et al. 2009. The Consumer Decision Journey. *McKinsey Quarterly*, www.mckinsey.com/business-functions/marketing-and-sales/our-insights/the-consumer-decision-journey (Accessed December 18, 2017).

Cunningham, Ceara Tess, et al. 2015. Exploring Physician Specialist Response Rates to Web-Based Surveys. *BMC Medical Research Methodology*. doi:10.1186/s12874-015-0016-z.

Deloite. 2016. Innovative Routes to Market Rethinking the Life Sciences Distribution Model, www2.deloitte.com/content/dam/Deloitte/uk/Documents/life-sciences-health-care/deloitte-uk-lshc-innovation-routes-to-market.pdf (Accessed January 9, 2018).

Denault, Jean-Francois. 2017. *The Handbook of Market Research for Life Science Companies: Finding the Answers You Need to Understand Your Market*, 1st edition, Productivity Press, Boca Raton, FL, 226 pages.

Derdenger, Timothy, & Kumar, Vineet. 2012. The Dynamic Effects of Bundling as a Product Strategy. *Harvard Business School*, https://hbswk.hbs.edu/item/the-dynamic-effects-of-bundling-as-a-product-strategy (Accessed January 30, 2018).

Eyal, Nir. 2014. *Hooked: How to Build Habit-Forming Products*, Portfolio, Toronto, Canada.

Farris, Paul, et al. 2006. *Marketing Metrics: 50+ Metrics Every Executive Should Master*, FT Press, Upper Saddle River, NJ.

Fifield, Paul. 2007. *Marketing Strategy*, 3th edition, Routledge, London, 328 pages.

Fuld, Leonard. 2006. *The Secret Language of Competitive Intelligence*, Crown Business, Boca Raton, FL.

Kotler, Philip, & Keller, Kevin Lane. 2015. *Marketing Management*, 15th edition, Pearson Higher Education, Upper Saddle River, NJ.

Harvard Business Review on Marketing. 2001. *Ideas with Impacts*, HBR School Press, Boston, MA.

Mas-Machuca, Marta, et al. 2014. A Review of Forecasting Models for New Products. *Intangible Capital* 10(1): 1–25. http://dx.doi.org/10.3926/ic.482, www.intangiblecapital.org/index.php/ic/article/view/482/405 (Accessed March 22, 2018).

Nordhielm, Christie, & Dapena-Baron, Marta. 2014. *Marketing Management, The Big Picture*, Wiley, New York.

Pit, Sabrina Winona, et al. 2014. The Effectiveness of Recruitment Strategies on General Practitioner's Survey Response Rates—A Systematic Review. *BMC Medical Research Methodology*. doi:10.1186/1471-2288-14-76.

Porter, Michael. 1999. *Competitive Strategy—Techniques for Analysing Industries and Companies*, The Free Press, New York.

Quelch, John, & Harding, David. 1996. Brands versus Private Labels: Fighting to Win. *Harvard Business Review on Brand Management*, HBS Press, Boston, MA, pp. 23–50.

Rao, Akshay, et al. 2000. How to Fight a Price War. *Harvard Business Review*, https://hbr.org/2000/03/how-to-fight-a-price-war (Accessed January 30, 2018).

Sufrin, Carolyn, & Ross, Joseph. 2008. Pharmaceutical Industry Marketing: Understanding Its Impact on Women's Health. *Obstetrical & Gynecological Survey* 63(9): 585–596. doi:10.1097/OGX.0b013e31817f1585. PMID 18713478 (Accessed January 30, 2018).

Vogler, Sabine, et al. 2015. Pharmaceutical Pricing and Reimbursement Policies: Challenges beyond the Financial Crisis, Vienna, Austria, www.researchgate.net/publication/282619555_Challenges_and_opportunities_for_pharmaceutical_pricing_and_reimbursement_policies (Accessed January 9, 2018).

Weinberg, Gabriel, & Mares, Justin. 2015. *Traction: How Any Startup Can Achieve Explosive Customer Growth*, Portfolio, New York.

Index

A

AARRR! metric, 141–142
Acquisition (AARRR!), 141–142
actions (implementation plan), 147
Activations (AARRR!), 141–142
active secondary research, 18–22
activity responsibility
    (implementation plan), 147
advertising, 127
    guidelines directly to customers (B2C), 173
    materials, analyzing, 67
advertising guidelines directly to customers (B2C), 173
advertising-to-sales ratio, 160–161
allowances, 123–124
American Hospital Association, 20
Anderson, Eric, 143
Ansoff, Igor, 102–103, 142
attributes, customer purchasing, 50
average cost per unit, break-even analysis, 158
average price per unit, break-even analysis, 158
average retention cost, 162

B

bargaining power of buyers, 78
bargaining power of suppliers, 77–78
barriers to entry, 78–79
behavioral segmentation, 102
Bhanji, S. Price, 107, 142
Bing, 21
Biotechnology Innovation Organization, 20
Bizer, George, 143
Board Reader, 21
bounce rate (BR), 167
brainstorming, 130
branding, 133
branding decisions, products, 115–116
break-even analysis, 157–159
Bullseye Framework, 130–132
bundles, products, 181
business model, assessing, 34–39
buyers, bargaining power, 78

C

CAC (customer acquisition cost), 161–162
    digital marketing, 168

193

channels, promotional, 129–132
click through rate (CTR), 166
client ecosystem, 44–49
client evaluation, 133–134
client-patient relationship, 178
collecting data. *see* data collection
combination model, 38
commercialization in therapeutics, 35–36
communication channels, Bullseye Framework, 131–132
company self-regulation, 125
competitive advantage, 110–114
competitive array, 64–66
Competitive Strategy, Techniques for Analyzing Industries and Companies, 79, 90
competitive-based pricing, 121–122
competitor analysis, 57–68, 133–134
   competitive array, 64–66
   data collection, 66–68
   evaluation, 61
   goal, 59–60
   identifying competitors, 61
   key competitors, 61
   preparation, 58–60
   profiles, 62–64
complexity, control processes, 153
components to systems, 108
conferences, analyzing, 67
confidence, products, 117–118
consumer group backlash, pricing, 125
control elements (implementation plan), 147
control processes, 5, 151–154
   barriers, 153–154
   implementation, 152–154
conventional health-care relationship, 185
conversion metrics, digital marketing, 166–167
conversion rate (CVR), 167

corporate vision, assessing, 31–32
corporate web DNA, 21
cost leadership strategy, 111, 113
cost-based pricing, 121
cost-per-click (CPC), 166
costs
   assessing, 33
   control processes, 153
   differentiation focus, 113
Court, David, 44, 90
CPC (cost-per-click), 166
CRM (Customer Relationship Manager), 100
CTR (click through rate), 166
Cunningham, C.T., 16, 27
current capabilities, assessing, 32–33
customer acquisition cost (CAC), 161–162
   digital marketing, 168
customer analysis, 40
   attraction, 40
   building customer profiles, 54–55
   client ecosystem, 44–49
   converting attention, 40
   customer behavior, 55–56
   decision-making process, 40–44
   gaining attention, 40
   identifying customer behavior, 52–54
   Kano model, 50–52
   purchasing, 40
customer behavior, 55–56
   identifying, 52–54
customer profiles, building, 54–55
customer purchasing process, 42
Customer Relationship Manager (CRM), 100
customer trends, 107
customer value-based pricing, 122
customization, digital products, 180

D

data collection, 12–13
  in-depth interviews, 13–14
  financial data, 24
  focus groups, 14–16
  government data, 19
  market research, 3
  online surveys, 16
  print media, 19–20
  public company data, 19
  sales data, 24
  search engines, 20–22
  security, 183
  trade and industry groups, 20
decision-making process,
    customers, 40–44
deep dive versus miles-wide
    research, 10
demographic segmentation, 102
Denault, Jean-Francois, 26
Derdenger, Timothy, 142
differentiation strategy, 112–113
digital marketing, 128, 138–142
  driving traffic, 139–140
  metrics, 164–168
  selling products online, 140–141
digital products, marketing strategies,
    179–184
direct sales distribution, 135–136
diversification, 108–110, 113
drug delivery company, 38

E

early payment discounts, 123–124
ease, products, 117–118
economic factors, marketing strategy
    implementation, 150
economic model, 53
environmental analysis, 76–85
epithetical TLR2, 107
ethics, market research, 25–26
excitement attributes, customer
    purchasing, 51
external analysis, 4–5, 29, 40–56
  customer analysis, 40–56
external pressures, marketing
    strategy implementation, 149
Eyal, Nir, 142

F

features decisions, products, 115
Feedly.com, 22
financial data, 24
fixed costs, break-even analysis, 158
fluctuating demand and
    perishability, 176–177
focus groups, 14–16
focusing on communication
    channels, 132
forecasting, market, 73–76
freemium pricing, 123
Fuld, Leonard, 21, 27
fully integrated model, 35–36
funding, assessing, 33
funnel customer purchasing
    process, 41

G

geographic segmentation, 101–102
Google, 21
Google Alerts, 23
government data, 19
governmental role, purchasing
    process, 49
GPOs (group purchasing
    organizations), 48, 125–126

H

*Handbook of Market Research for
    Life, The,* 26
Harding, David, 44, 90

health care services, marketing strategies, 174–179
health-care digital products, marketing strategies, 179–184
Heaney, Paul, 143
Hemmelgarn, B., 16, 27
Hill, Pauline, 57–58
Hook Model, 119–120
hospitals, 48
Hugues, Gwyndaf, 143
human resources, assessing, 33

I

ICE score, 117–119
impact, products, 117–118
implementation, 145–151
  barriers, 149–151
  control processes, 152–154
  planning, 147–148
  versus strategy, 146–147
incentives, marketing strategy implementation, 151
in-depth interviews, 8, 13–14
industry groups, 20
industry rivalry, 79–80
inside-out perspective, 60
insurers, 47
intangibility, 175–176
integrated marketing, 100
intellectual property (IP), assessing, 33
internal analysis, 4, 29, 30–40
  assessing business model, 34–39
  assessing corporate vision and mission objectives, 31–32
  assessing current capabilities, 32–33
internal marketing, 100–101
internal pressures, marketing strategy implementation, 150

interviews, in-depth, 13–14
IP (intellectual property), assessing, 33

J

Journal of Medical Marketing: Device, Diagnostic and Pharmaceutical Marketing, The, 16, 27
JustADAPTE, 107
JustBIO, 107

K

Kano model, 42, 102
Kelly, D., 16, 27
key competitors, identifying, 61
key opinion leaders (KOLs), 107
key product decisions, 115–116
Keyhole, 23
keyword strategy, digital marketing, 140
KOLs (key opinion leaders), 107
Kumar, Vineet, 142

L

lab techs, typical, 56
lack of incentives, marketing strategy implementation, 151
leadership barriers, marketing strategy implementation, 150
leads, 75
learning model, 53
legal factors, marketing strategy implementation, 149
liability, digital products, 183
life sciences, marketing strategies, 171–174
life sciences purchasing process, relationships, 46
lifetime customer value, 162–164

Loftus, Peter, 143
low-end disruptive innovation strategy, 111

M

management barriers, marketing strategy implementation, 150
management by exceptions, control processes, 153
Mares, Justin, 130
market analysis, 68
   market forecasting, 73–76
   TAM-SAM-SOM model, 68–72
market development, 105–107, 113
market forecasting, 73–76
market penetration, 103–104, 113
market research, 2–3, 7–8
   active secondary research, 18–22
   data collection, 3, 12–13
      in-depth interviews, 13–14
      focus groups, 14–16
      online surveys, 16
   ethics, 25–26
   miles-wide versus deep dive research, 10
   passive secondary research, 18–22
   preparing plan, 11–12
   primary and secondary, 8
   quantitative and qualitative data, 8–10
market segmentation, 101–102
market size estimation, 68–72
marketing
   integrated, 100
   internal, 100–101
   relationship, 99–100
marketing metrics, 155, 168–169
   advertising-to-sales ratio, 160–161
   average retention cost, 162
   benefits, 156
   customer acquisition cost, 161–162
   digital marketing, 164–168
   lifetime customer value, 162–164
   marketing percentage of CAC, 161–162
   pre-revenue ratios, 156–159
   ROS (return on sales), 160
marketing mix, developing, 114–138
marketing models
   choosing, 95–99
   product-focused, 97
   production model, 96–97
   selling, 97–98
marketing percentage of CAC, 161–162
marketing strategies, 1
   based on products and market, 102–110
   competitive advantage, 110–114
   control, 145, 151–154
   developing, 5
   developing marketing mix, 114–138
   digital, 138–142
   distribution, 132–138
   health care services, 174–179
   health-care digital products, 179–184
   implementation and control systems, 5
   implementation, 145–151
   importance, 2
   life sciences, 171–174
   marketing models, choosing, 95–99
   market-segmentation, 101–102
   planning, market research, 2–3
   Porter, Michael, 110
   pricing, 121–126
   promotional, 126–132
   selecting vision, 91–95
   situation analysis, 4–5
Marusic, Lidija, 68
McClure, Dave, 141, 144
media backlash, pricing, 125

Medical Device Manufactures Association, 20
MedLung's SAM, 71
messaging, promotional, 129–132
metrics. *see* marketing metrics
miles-wide versus deep dive research, 10
mission objectives
　assessing, 31–32
　building, 31–32
monitoring, control processes, 153
Moreno, Katherine Parra, 2, 49
multiple revenue strategies, 180

N

Nanopore Diagnostics, 106–107
NetVibes, 22
neutral attributes, customer purchasing, 51
new-market innovative disrupters, 111
niche markets, 105–106
no research/development only model, 38
non-retail sales distribution, 135

O

off-peak pricing, 177
O'Hear, Steve, 177, 184
online media, analyzing, 67
online sells, 140–141
opportunity, identifying, 57–58
organizational model, marketing strategy implementation, 151
organizational segmentation, 102
outside-in perspective, 60

P

packaging decisions, products, 115
page views (digital marketing), 165
paid traffic, monitoring, 166
patent databases, analyzing, 67
patients, 46
performance attributes, customer purchasing, 50–51
perishability, 176–177
personal selling, 128
PEST (political, environmental, social, and technology) model, 77
PhRMA, 20
physical resources, assessing, 33
Pit, S.W., 27
plan (market research), preparing, 11–12
planning, implementation, 147–148
platform companies, 37
PMF (Product-Market Fit), 116–117
political, environmental, social, and technology (PEST) model, 77
political factors, marketing strategy implementation, 150
Porter, Michael, 79, 90, 110
Porter's Five Forces, 5, 40, 77–80
　bargaining power of buyers, 78
　bargaining power of suppliers, 77–78
　barriers to entry, 78–79
　industry rivalry, 79–80
　threats of substitutes, 79
positioning, 133
PR (public relations), 128
pre-revenue ratios, 156–159
price adjustments, 104
price differentiation, 107–108
price discounts and allowances, 123–124
price endings, 124
price skimming, 122–123
pricing, off-peak, 177
pricing in pharmaceutical products, 124–125
pricing strategies, 121–126

primary caregivers, 47, 48
primary market research, 8, 13–16
print media, data collection, 19–20
prioritizing prospects, 132
private company websites, analyzing, 66
private insurers, 47
Problem-Solution Fit, 116
product, assessing, 33
product bundling, 108
product decisions, key decisions, 115–116
product development strategies, 107–108, 113
product diversification, 108–110, 113
product service, definition, 60–61
product-focused model, 97
production model, 96–97
Product-Market Fit (PMF), measuring, 116–117
Product/Market matrix, 102–103
products
  branding, 133
  bundles, 181
  confidence, 117–118
  digital, marketing strategies, 179–184
  discovery, 183–184
  ease, 117–118
  impact, 117–118
  positioning, 133
  versus services, 174–175
  trial periods, 181–182
  unbundling features, 181
profiles, competitor, 62–64
promotional messages and channels, 129–132
promotional strategies, 126–132
psychoanalytical model, 53
psychographic segmentation, 102
psychological model, 53
public company data, 19

public insurers, 47
public relations (PR), 128
purchasing, 40
purchasing process, relationships, 46
Pyakurel, S., 16, 27

Q

qualified leads, 75
qualitative data, 9–10
quality decisions, products, 115
quality discounts, 123–124
Quan, H., 16, 27
quantitative data, 8–9
Quelch, John, 44, 90

R

ranking and classification of promotional channels, 130
Rao, Akshay, 143
recovery model, 38
Referral (AARRR!), 141–142
realistic assessments, 72
relationship marketing, 99–100
relationships
  client-patient, 178
  conventional health-care, 185
  life sciences purchasing process, 46
Request for Proposals (RFPs), 125–126
Requests for Quotations (RFQs)., 125–126
research. *see* market research
research model, 37
resources, marketing strategy implementation, 151
resources (implementation plan), 147
retail sales distribution, 135
Retention (AARRR!), 141–142
return on investment (ROI), digital marketing, 168

return on sales (ROS), 160
Revenue (AARRR!), 141–142
revenue metrics, digital marketing, 168
reverse attributes, customer purchasing, 51
RFPs (Request for Proposals), 125–126
RFQs (Requests for Quotations), 125–126
Rich Site Summary (RSS), 22–23
Rivière, Marc, 30, 72
ROI (return on investment), digital marketing, 168
ROS (return on sales), 160
RSS (Rich Site Summary) feeds, 22–23
RSSOwl, 22
Rupert, E, 16, 27

S

sales data, 24
sales force coverage, 156–157
sales promotions, 127–128
Schindler, Robert, 143
search engines, 20–22
secondary market research, 8, 17–18
  active, 18–22
  passive, 18–22
Secret Language of Competitive Intelligence, The, 21, 27
security, data collection, 183
segmentation, 101–102
self-health sector, growth, 186
services versus products, 174–175
situation analysis, 4–5, 29–30
  competitor analysis, 57–76
  environmental analysis, 76–85
  external analysis, 40–56
    customer analysis, 40–56
  internal analysis, 30–39, 39–40
    assessing business model, 34–39

assessing corporate vision and mission objectives, 31–32
assessing current capabilities, 32–33
SLEPT (Social, Legal, Economic, Political, and Technological), 5, 40, 77, 80
  building, 82–83
  Economic, 82, 84
  Legal, 81–82, 83
  Political, 82, 84
  Social, 80–81, 83
  Technology, 82, 84–85
Slide Share, 21
Social, Legal, Economic, Political, and Technology (SLEPT). *see* SLEPT (Social, Legal, Economic, Political, and Technology)
social factors, marketing strategy implementation, 149
social media, digital marketing, 140
social media tracking, 23
sociological model, 53
strategic adaptability, 95
strategic commitment, 92–93
strategic objective (implementation plan), 147
strategic opportunism, 93–94
strategies. *see* marketing strategies
strategy, versus implementation, 146–147
Sufrin, Carolyn, 143
suppliers, bargaining power, 77–78
SWOT model, 85–88
  opportunity, 87
  strategy applications, 88–90
  strengths, 87
  threats, 87–88
  weaknesses, 87
systems, components, 108

T

TAM-SAM-SOM model, 68–72
target market, determining, 101–102
targets, control processes, 153
technological factors, marketing strategy implementation, 150
technological innovation, 107–108
testing communication channels, 132
therapeutics, commercialization, 35–36
Theronas, 186
threats of substitutes, 79
threshold attributes, customer purchasing, 50
timetable (implementation plan), 147
TOWS matrix, 88–90
Traction Bullseye Framework, 130–132
trade groups, data collection, 20
traffic, digital marketing, 139–140
traffic metrics, 164–167
trial periods, products, 181–182
Twitter, tracking, 23

U

unbundling features, products, 181
uniformity of service, 178
unique selling proposition (USP), 115
unique visitors (digital marketing), 165
unmet needs, 105
USP (unique selling proposition), 115

V

value-added reseller (VAR), 134–136
VAR (value-added reseller), 134–136
virtual model, 37
visibility, increasing, 104
vision, 31–32
   selecting, 91–95
   strategic adaptability, 95
   strategic commitment, 92–93
   strategic opportunism, 93–94
Vo, T., 16, 27
Vogler, Sabine, 143
voice-of-the-customer initiative, 56

W

Wadha, H., 17, 27
Warble, 23
web surveys, 8
web traffic sources, monitoring, 165–166
website optimization, digital marketing, 140
Weinberg, Gabriel, 130, 143